SUPERBUGS

SUPERBUGS

An Arms Race against Bacteria

WILLIAM HALL, ANTHONY MCDONNELL, AND JIM O'NEILL

Harvard University Press

Cambridge, Massachusetts

London, England

2018

First Printing

Library of Congress Cataloging-in-Publication Data

Names: Hall, William (Author of Superbugs), author. | McDonnell, Anthony
  (Co-author of Superbugs), author. | O'Neill, Jim (Economist), author.
Title: Superbugs : an arms race against bacteria / William Hall, Anthony
  McDonnell, and Jim O'Neill.
Description: Cambridge, Massachusetts : Harvard University Press, 2018. |
  Includes bibliographical references and index.
Identifiers: LCCN 2017049342 | ISBN 9780674975989 (alk. paper)
Subjects: LCSH: Drug resistance in microorganisms. | Bacterial
  diseases—Prevention. | Medical policy.
Classification: LCC QR177 .H35 2018 | DDC 616.9/041—dc23
LC record available at https://lccn.loc.gov/2017049342

Contents

Foreword

SALLY DAVIES

A s a doctor, I had been aware of the problem of antimicrobial resistance and drug-resistant infections for many years, but I did not begin to fully understand the rising global scale of the challenge until, in my role as chief medical officer for England, I prepared an annual report on antimicrobial resistance in 2013. Later that year I published a book called *The Drugs Don't Work: A Global Threat* to raise awareness of this scientific problem among a wider audience, and I pressed the then prime minister, David Cameron, to commission an independent Review on Antimicrobial Resistance. We needed a thorough overview that would analyze the complex economic aspects of the issue and garner broader international attention. The Review, which was established in 2014 and chaired by Jim O'Neill, was given the task of examining the economic and policy aspects of the problem of antimicrobial resistance, as well as outlining the scope of the problem and making recommendations for a global action plan. Its final report was submitted in May 2016.

Society has taken antibiotics for granted for far too long. Over the past seventy years, they have transformed our life expectancy and our lives. Because of inappropriate use, however, in both humans and animals, drug-resistant infections have emerged and spread around the world, and existing antibiotics have become increasingly ineffective. At the same time, we have not developed new antibiotics or alternative treatments to replace them. If we do not take action now, minor injuries could be life-threatening, and the advances of modern medicine will be under threat. We are already seeing the impact of drug-resistant infections on the young, the old, pregnant women, and those with compromised immune systems, and this situation will only get worse. Cancer and its treatments generally result in patients becoming more susceptible to infection, which becomes a real problem if we do not have effective antibiotics. Although the risks of infection associated with childbirth have fallen considerably in most countries over recent decades, without effective antibiotics the incidence of these infections could again dramatically increase, in both mothers and children. We also rely on antibiotics to prevent life-threatening infections following many relatively routine surgeries, such as hip replacements and organ transplants. Drug-resistant infections threaten not only human health, but also agricultural livelihoods and food security, and therefore, the whole economy.

While drug-resistant infections are a major killer already, with many hundreds of thousands of people across the world dying every year, a very large number of people die from not having access to antibiotics in the first place. Those who need antibiotics should receive appropriate treatment; generic antibiotics must still be produced, supplied, and distributed globally and affordably. But it is essential to reduce excessive use. "Access not excess" is a simple but important mantra we can use when thinking about this complex problem.

The Review on Antimicrobial Resistance was a great success, and, alongside a drive from the UK government and many others from the public, private, and philanthropic sectors, it helped to galvanize international action at meetings of both the G20 (Group of Twenty developed and emerging economies) and the United Nations General Assembly in 2016. These forums have raised awareness of the issue to the highest level, but we must recognize it is just the beginning. I will continue to do all that I can to achieve progress, including through the newly established ad hoc

UN Interagency Coordination Group on Antimicrobial Resistance, of which I am a convener.

This book is a vital contribution that clearly articulates the case for action, and, crucially, how to do it. At the core of addressing drug-resistant infections are a number of economic challenges which must be surmounted. *Superbugs* has the promise to bring these economic policy solutions to a wider international audience, raising public awareness and helping to stimulate action. It also makes a convincing case for policymakers to take the next steps, and contends that we have reason to be optimistic if we are decisive now and address this complex problem with sufficient resources. We must act soon—otherwise this terrible problem will continue to worsen, and we will leave the world in a much poorer state for our children and grandchildren to come.

Introduction

Bacteria live inside all of us. On average, every human body contains 100 trillion bacterial cells—three bacteria for every human cell. If all of these bacteria were removed, we would each lose about 2 kg. (4.5 lbs.) in weight. Thankfully, most of these bacteria are either benign or play useful roles, such as helping us digest food. However, some of them that might be harmless in our nose or gut can be lethal in our lungs or bloodstream. Other bacteria that are not normal inhabitants of the body can also infect us, causing diseases such as tuberculosis and diphtheria. For much of human history, these kinds of bacterial infections were some of the world's biggest killers, claiming huge numbers of lives.

During the late nineteenth and early twentieth centuries, we became much better at preventing infections, but doctors still remained powerless to treat them after they started. This changed in the 1940s when penicillin, the world's first modern antibiotic, became available. Patients who had been near death were able to return to full fitness just days after taking what was often called a miracle cure.

But the miracle did not last. Unlike the situation in almost every other area of medicine, antibiotics became less effective the more they were used. In response to the onslaught of penicillin and other antibiotics, drug-resistant bacteria—bacteria with characteristics that protect them from attack—survived and thrived. Bacteria have now become resistant to many of the drugs in our current arsenal, and fewer new drugs are being produced.

We have reached the point where scientists are warning that we face one of the greatest threats to humankind: not only could many common infections become incurable, but a lack of effective antibiotics could effectively end much of modern medicine, since we rely on these drugs to keep us healthy during surgery and cancer treatment. A sobering example from 2016 provides a glimpse into a possible future: a woman in a hospital in the United States had a bacterial infection in her leg that was completely resistant to every available antibiotic—twenty-six in all. Nothing further could be done to treat the infection. Despite having access to one of the best health-care systems in the world, the woman eventually died.

In 2014, Sally Davies, chief medical officer for England and author of this book's foreword, convinced Prime Minister David Cameron that the United Kingdom needed to take a leadership role on the issue. Cameron commissioned an economic Review on Antimicrobial Resistance to look into the problem, choosing as chair the economist Jim O'Neill, one of the authors of this book. The other authors, William Hall and Anthony McDonnell, were also members of the Review team: Hall was the senior policy adviser, and McDonnell was head of economic research. As economists and policymakers outside of the scientific world, we knew relatively little about this problem when we first began work on it. Working on the Review we quickly realized that to make any progress, it was essential to reach out beyond the medical and scientific world: the issue needed to be analyzed and explained from an economic perspective. The health ministries of the world got the problem. The finance ministries, in general, did not. Finance ministers wanted to know the answers to questions such as these: How many people currently die as a result of antimicrobial resistance, and how many more would die if we took no action? What are the economic costs both today and in the future? What policy interventions do we need to make, and how much would they cost? Who would pay for them? These were some of the questions the Review on Antimicrobial Resistance set out to answer, and in this book, we outline the solutions to this problem through a policy and economic lens.

We should mention one issue about terminology before we go any further. Note that the title of the Review includes the word *antimicrobial*, not *antibacterial*. *Microbe* is a broad term that includes not only bacteria but also other pathogens: viruses, fungi, and parasites (which cause diseases like HIV, yeast infections, and malaria, respectively). The problem of drug resistance affects diseases caused by all of these kinds of microbes—hence the umbrella term "antimicrobial resistance." But in this book we focus primarily on resistance to antibiotics that has developed in bacteria, since the problem has been known about for decades in the scientific community, but there has been insufficient international action to counter it. It is also a truly global problem that has the potential to affect every person on the planet.

Many people see the main challenge for stopping drug-resistant infections as developing new antibiotics. However, this is but one part of the solution. In order to win this arms race against bacteria, we need not only to improve our tools, but also to reduce unnecessary use, so that the drugs we have will remain effective for a longer time. Antibiotics are being overused in both people and animals. Too often, people take antibiotics when they are not needed. One study from 2013 estimated that of the 40 million people who are given antibiotics for respiratory problems in the United States every year, around 27 million do not need them. When an antibiotic is prescribed in such a way that it saves a life, the risk of increased resistance is a small price to pay. But we should not be taking this risk in cases where we gain little or no curative benefit. Our farming and aquaculture systems also use excessive amounts of antibiotics—often similar drugs to those used in humans. Residues of antibiotics enter the environment in countries where antibiotics are manufactured. This flood of antibiotics drives resistance and makes our drugs less effective. We do need new drugs—but unless we use them better, they are unlikely to be useful for long.

Antimicrobial resistance is often described as one of the biggest threats to humankind, but we believe that the doomsday scenarios can be averted. We can change how we buy medicines and organize health care in ways that can make a huge difference quickly. Increased awareness on the part of both the public and health-care professionals can play an important role in improving the situation. A recent study by the Wellcome Trust in the United Kingdom, for example, found that most people do not know that taking antibiotics can lead to the development of drug-resistant bacteria, and if they have heard of resistance, many think that it is their body that becomes resistant to the drug, rather than the bacteria.

We also need to get the word out that the problem is an urgent one today, not something that is just looming on the horizon. In response to the final report of the Review on Antimicrobial Resistance, Mario Monti, former prime minister of Italy, said, "Antimicrobial resistance, an insidious and particularly devastating phenomenon, is definitely worth the urgent attention of global leaders. But it is unlikely to actually reach the leaders' agenda as a key priority unless it is pushed up there by a compelling exercise of awareness-raising, policy-advocacy and coalition-building, strong enough to overcome the short-term bias of current politics." Indeed, as we demonstrate in this book, approximately 1.5 million people die every year already as a result of antimicrobial resistance. That is a huge number. It is about the same as the number of annual deaths from diabetes, and more than the number of deaths from traffic accidents.

Such large numbers can be hard to imagine, and so in this book, in addition to interviewing leading politicians and international experts on antimicrobial resistance, we interview people directly affected by it—people who have lost loved ones, and those who have spent large parts of their lives treating patients and seeing the problem worsen over time.

At its heart, antimicrobial resistance is a problem of economic and political failure. We need pharmaceutical companies to develop new antibiotics, but we must at the same time limit their use, which then makes it difficult for companies to recoup their investment. The current rewards-based system does not work. In this book we combine economic policy ideas with our experience of global politics and global business investment to present workable solutions that policymakers can implement.

The book is divided into two parts. Part I outlines the nature of the problem and its causes. In Chapter 1, we discuss how antibiotics became one of the most important discoveries in human history, and how they made possible the emergence of modern medicine. We then describe, in Chapter 2, how the rise of drug resistance rendered antibiotics less effective. We present data showing just how expensive a problem it is and will be in the future, in terms of both lives lost and dollars spent. Chapter 3 reviews the economic and political factors that are behind our failure to tackle the growing threat of drug resistance.

Part II offers a comprehensive set of solutions for antimicrobial resistance, including both how to encourage the development of new antibiotics and how to reduce their use. Chapter 4 presents our proposal for improving incentives for companies to create new antibiotics. Then we turn to prevention. As explained in Chapter 5, we can keep people healthy by getting more people to wash their hands, building sanitation systems, improving hospital cleanliness, and increasing the use and development of vaccines. These initiatives will not only keep people from getting sick, but will reduce the development of antibiotic resistance. Chapter 6 focuses on ways to reduce the unnecessary use of antibiotics through methods such as developing rapid diagnostic devices and improving global disease surveillance. We then turn to agriculture and the environment in Chapter 7, where we make suggestions for how to limit antibiotic use in animals that are raised for food, as well as encouraging manufacturers to curtail discharge of antibiotic waste. We conclude by discussing the political steps required to achieve further international action.

Antimicrobial resistance is a global, multifaceted problem. The cost, both in economic terms and in human lives lost, is huge today. It will become far higher if we do not act fast. However, we can make headway by establishing the right economic and social incentives. Such large-scale incentives will require global collaboration, commitment, and expertise. The first steps have already been taken following the Review on Antimicrobial Resistance and a declaration at the United Nations in 2016. But in many areas there remains a need for real action to follow these strong words. Now is the time to make the investments and implement the solutions to ensure that we prevent this terrible health problem from spiraling into a global catastrophe.

1

The Problem of Drug Resistance

When a Scratch Could Kill

For many readers of this book, particularly those who live in high-income countries and were born after the Second World War, the fear of infectious diseases may not be a source of anxiety. Diseases like Ebola or Zika, both caused by viruses, still capture headlines. But in high-income countries, people's experiences of infectious illnesses consist mostly of occasional bouts of the common cold and other self-limiting or easily treatable infections. The perceived security—some would say complacency—that we enjoy is the product of growing up in an era when many of us enjoy ready access to effective antibiotics, the protection afforded by good health care, and high standards of hygiene in public and in the home.

Of course, infectious diseases—both long-established ones, such as tuberculosis, and more modern threats, such as AIDS—remain real and pervasive threats in many parts of the world, particularly in low- and middle-income countries that lack a reliable public health infrastructure and access to essential medicines. But for most of the developed world, patterns of disease have shifted profoundly over the course of the past century.

In this chapter, we explore the extent to which the treatment of infectious diseases in Europe and North America has changed during the past two centuries. We consider how a series of breakthroughs in scientific research and understanding—beginning with the development of the "germ theory" in the late nineteenth century and culminating in the discovery and development of effective modern antibiotics during the first half of the twentieth century—have shaped the world we live in today and laid the foundations for the crisis we face, as we find ourselves on the cusp of what Margaret Chan, former director general of the World Health Organization, has described as a "postantibiotic era."

The Pre-Antibiotic Era: Experiences and Understanding of Infection

In developed, high-income countries today, the burden of infectious disease—in terms of both disease incidence and mortality—is low. A typical adult in the United States is more likely to die of accidental or violent causes than of any type of infectious disease. In the world as a whole, noncommunicable conditions (such as cancers and heart disease) cause four times as many deaths as infectious diseases. An American born in 2015 can expect to live to the age of nearly eighty, while a typical sixty-five-year-old baby boomer now approaching retirement age can expect to enjoy another nineteen years of life.

These statistics are in startling contrast to the life expectancy, patterns of illness, and causes of mortality in western Europe and North America during the nineteenth and early twentieth centuries. For example, in 1841, average life expectancy in England (then one of the most prosperous countries on earth) was just forty-one years. Within that country's major cities life expectancy was even lower—a child born in Manchester faced a meagre life expectancy of twenty-five years. By 1900, life expectancy at birth was only forty-seven years, and in the United States, approximately a third of all deaths were from tuberculosis, pneumonia, or gastroenteritis.[1] At the beginning of the twentieth century in the United States, one infant in ten died before their first birthday. For every thousand live births, between six and nine mothers died during or shortly after childbirth. Sepsis—bacterial infection of the bloodstream—was responsible for forty percent of these deaths.

Following the industrial revolution that took place in the United Kingdom and many parts of western Europe and North America in the nineteenth century, urban populations grew quickly, often resulting in squalid and overcrowded living conditions. In dense and unsanitary living quarters, with limited sewerage and access only to shared—and often polluted—drinking water sources, air- and waterborne diseases spread easily. The poor suffered most.

However, affluence and status could not completely protect people against infectious disease. An example from American presidential history concerns President Calvin Coolidge's family. On a hot afternoon in 1924, just a few days before the annual July Fourth celebrations, Coolidge's two sons, John and Calvin Jr., spent the afternoon playing tennis on the White House grounds. The younger of the two, sixteen-year-old Calvin Jr., began to suffer pain from a blister on his toe, probably as a result of wearing his tennis shoes without socks. The blister became infected. Calvin Jr. developed a fever, and his condition rapidly deteriorated over the next few days. He was transferred to the Walter Reed Medical Center, suffering from *Staphylococcus aureus* blood poisoning. The president wrote in a letter to his father on July 4, "Calvin is very sick. . . . Of course he has all that medical science can give but he may have a long sickness with ulcers, then again he may be better in a few days." Sadly, the teenager died on July 7, just a week after his game of tennis. Even privilege and access to the very best medical care of the day could offer no defense against death from an injury and subsequent infection.

This poignant story is just one of many notable examples that illustrate the reality of disease in the pre-antibiotic era: infectious diseases were a blight across all sections of society, even affecting the affluent and otherwise young and healthy. A seemingly innocuous scratch or cut truly could kill.

During the nineteenth century, tuberculosis (TB), also called consumption, or the "white plague," was a pervasive affliction. TB was common across all sections of society, especially in dense urban areas. In major European cities such as London, Paris, and Stockholm, annual mortality rates from TB were 800–1,000 deaths per 100,000. The disease accounted for 40 percent of deaths among the urban poor, with latent TB infection rates estimated at between 70 and 100 percent in some urban areas. The death rate among those who developed active TB infections was 80 percent.

The best efforts to treat TB were offered by the so-called sanatorium movement, whose proponents contended that the illness could be cured through relocation to places where the air was cleaner and circulated more freely. An extended stay at a sanatorium—typically located in a mountainous, rural, or seaside location—was often prescribed for more affluent TB patients, but in reality it offered little more than symptomatic relief from the disease. Nonetheless, considerable faith was placed in the curative ability of sanatoria, even after the discovery of the bacterium that causes TB (described later in this chapter). Although experimental treatments based on new discoveries emerged in the early 1900s, the recovery rates from TB remained persistently low, and the majority of patients "treated" in sanatoria were dead within five years of their discharge.

One of the authors' own family members, Joyce Pickard spent three and a half years of her childhood in a British hospital that specialized in the treatment of children afflicted with nonpulmonary TB (see Figure 1.1). She lost the ability to walk and became permanently disabled by a severe TB infection in her hip bones. From late 1937 to the summer of 1941, she was almost

Fig. 1.1. A British hospital for children with tuberculosis in the 1930s. Credit: "Sun Therapy at Alton Hospital," Wellcome Collection (CC BY 4.0).

completely confined to her bed in a hospital on the south coast of England. More than two hundred miles from her home in Yorkshire, she had little exposure to the outside world. Without antibiotics there was no way to treat the underlying infection effectively; her treatment consisted of isolation and fresh air. Her bed would be wheeled outdoors even in the depths of winter, when she recalled that she would develop chilblains on her hands from the cold, and doctors would do their ward rounds while shivering in heavy overcoats. Such a protracted period of treatment, and the confinement it involved, seems extraordinary to us today, particularly since the disease is now both preventable and treatable.

This picture of infectious disease has changed almost entirely thanks to a series of breakthroughs in our understanding of disease, and subsequently in the discovery of antibiotics to treat them. Starting in the mid-nineteenth century, average life expectancy has increased by a greater amount, and more rapidly, than at any other period in human history, and this increase is attributable to the decline in infectious disease. But with the rise in antibiotic resistance potentially rendering vital treatments ineffective, are we facing a postantibiotic era that resembles this pre-antibiotic past?

The Development of the Germ Theory

We take it for granted today that—with a few exceptions—the causes of infectious diseases are well understood. These illnesses are spread from person to person, and in the environment around us, by a pathogen of one sort or another, of which bacteria and viruses are the two most common. Bacteria are single-celled organisms that live within us and around us in vast quantities, usually benignly coexisting with us (and even helping to keep us healthy), but in certain circumstances colonizing our bodies, multiplying uncontrollably, and causing illness. Viruses, meanwhile, are particles around a hundred times smaller than bacteria that penetrate our own cells and multiply within to cause diseases ranging from the common cold and influenza to rabies and Ebola.

But these facts have not always been known, and it was only a series of significant breakthroughs during the late nineteenth and early twentieth centuries that allowed scientists to properly understand the causes of many infectious diseases. Before the middle of the nineteenth century, people recognized that diseases like cholera were contagious—that is, they spread within a population—but the cause of the contagion was unknown. Many theories emphasized the role of "miasmas," foul-smelling vapors that were thought to transmit disease, exacerbated by environmental conditions and proximity to "filth." Such theories did account for certain key characteristics of the diseases—in the case of cholera, for example, areas hit by outbreaks would be dogged by foul stenches and conditions, and disease outbreaks were more likely to occur during hot summer months than in the cold of winter. However, the source of the contagion was often misidentified. For instance, conventional wisdom blamed typhoid on fecal contamination and poor personal hygiene, but little attention was paid to the role of contaminated drinking water or food. Another common hypothesis stated that some kind of predisposition—whether by birth or acquired through an individual's environment and circumstances—was crucial to the development of the disease, downplaying the importance of exposure to the infection and how it occurs.

All of these ideas changed with the emergence of the germ theory, which posited that living microorganisms cause infectious diseases.

One of the founders of the germ theory was Louis Pasteur, a chemist who taught in Lille, in northern France, in the mid-nineteenth century. Lille was in an agricultural region and was the center for the industrial fermentation of sugar beets into alcohol. Pasteur became interested in the process of fermentation, and by means of a high-resolution microscope he was able to observe and identify the different microorganisms responsible for fermentation and putrefaction in food. He realized that microorganisms got into food through contamination from the environment, not, as had been thought, through a spontaneous decay process. This discovery highlighted the fact that microorganisms originated from the outside, not the inside. Could the same process—exposure to microorganisms—be responsible for certain diseases?

The other founder of germ theory is Robert Koch, a German physician and biologist. In the late 1870s, Koch published his investigations into bacteria as disease-causing microorganisms. He demonstrated the process by isolating pathogens from diseased animals and inoculating the pathogens into healthy animals, showing how diseases could be transmitted. In 1882 Koch successfully isolated the tuberculin bacillus, which he claimed to be the cause of all tuberculosis illness. This marked a significant step forward in the understanding of the cause and mechanism of transmission of TB.

In 1854, around the same time as Pasteur's early work on microorganisms, a London-based physician named John Snow did pioneering work on cholera that elucidated the source of a major cholera outbreak in London's Soho district. Snow rigorously tracked and mapped cases as the outbreak emerged and demonstrated that a cluster of cases occurred around a particular water source—a shared pump. The data supported his theory that cholera was transmitted in drinking water. He staged a stark intervention by persuading parish authorities to remove the pump handle to take it out of service, an act that slowed the spread of the outbreak. His action provided compelling circumstantial evidence that the water source had played a role in the outbreak, establishing Snow's position as the founder of infectious disease epidemiology.

Another important innovator was Joseph Lister, a British surgeon. Lister became interested in the implications of Pasteur's findings for patients with wounds that became infected after surgery. Drawing on Pasteur's conclusions that putrefaction was caused by contamination with microorganisms from the environment and the open air, Lister began experimenting in the 1860s with using antiseptics to clean and souse wounds during surgery, producing sharp reductions in postoperative infections. Lister's work was one of the first instances of the application of Pasteur's ideas to human medicine.

Early Interventions to Reduce the Infectious Disease Burden

These discoveries, and the growing acceptance of the germ theory, played a crucial role in attempts to prevent and treat infectious disease during the second half of the nineteenth century. Most notable were what Sally Davies, chief medical officer for

England, and others have categorized as the "first wave" of recognizable public health interventions—programs offering basic interventions to protect the population's health, including vaccinations and the construction of effective sanitation infrastructure. The introduction of proper sanitation in Europe's growing cities—exemplified by the construction of London's first modern sewage system during the 1860s—is one of the most significant of these interventions, offering increased protection from water-borne risks. Similarly, interventions such as vaccination programs and the pasteurization of milk became increasingly common during the second half of the century, directly driven by the work of Pasteur and other originators of the germ theory.

Although the effect of these interventions was significant, some scholars (especially Thomas McKeown) have argued that improvements in health had already emerged during the first half of the nineteenth century, and that the most important factor in this improvement was the rising affluence of the United States and western Europe, and associated improvements in living conditions and nutrition. In any case, it appears that the greatest gains in population health were appearing before the advent of effective antimicrobial drugs, thanks to large-scale interventions to reduce people's exposure to the agents that were now known to be causing disease.

The Birth of the Antibiotic Era

With the emerging understanding of germ theory and the nature of infectious diseases, there was a shift in medical treatment. As the role of microbes in disease gained currency, so too did the notion of a medicinal "magic bullet," a drug that could selectively target a pathogen and cure disease.

Early Breakthroughs in the Hunt for a "Magic Bullet"

One of the first breakthroughs in the search for a magic bullet developed out of an exploration of toxic substances that could be used to target microbes, such as the one that causes syphilis. Like Pasteur, many of the scientists involved in this research were working in industrial environments.

One of the pioneers in these efforts was Paul Ehrlich, a German chemist who was studying how different types of chemical dye could selectively stain different tissue types. Ehrlich also noted that some arsenic-based compounds targeted and killed certain pathogens. This observation led him to the concept of the magic bullet, a chemical that would destroy a specific pathogen in an organism without harming the organism itself. The causal agent of syphilis, a kind of bacterium called a spirochete, had been identified in 1905, and Ehrlich and his colleague Sahachiro Hata set to work testing hundreds of compounds, trying to find one that would destroy it. The breakthrough came in 1907, with the 606th compound that Hata tested. This compound, later renamed arsphenamine, seemed to be remarkably active against certain bacteria, including the one that causes syphilis. The compound was brought to market in 1910 as Salvarsan. This product, and its later derivatives, remained the standard treatment for syphilis until the Second World War, and is regarded as the first true modern antimicrobial drug to enter common usage.

The success of Salvarsan was a catalyst for further explorations by like-minded chemists. Most notable among these was the discovery in 1932 by scientists at the German company Bayer of sulfonamides, or sulfa drugs, a new class of antimicrobial drugs. These were the first true synthetic "antibiotics" to come to market, proving to be a breakthrough in treatment if not a commercial success for the company that had discovered them.[2]

These early advances were built directly on the foundations laid by Koch, Pasteur, and others, and firmly validated Ehrlich's early hypothesis about the notion of medicinal magic bullets. Further research grew out of the expertise of scientists working within the German chemical industry on synthetic chemical compounds. But the most significant discovery of the century was to come from an entirely separate line of research, and it came about as the result of a serendipitous accident.

The Fortuitous Discovery of Penicillin

As Ehrlich's and the Bayer company's research efforts were progressing in Germany, a researcher at St. Mary's Hospital in London, Alexander Fleming, was undertaking research into a group of bacteria called staphylococci. His research might never have entered the history books were it not for a series of extraordinary, fortuitous coincidences during the summer of 1928, which led to a discovery that would define his career—and change modern medicine.

When departing on his summer break in August, Fleming left his laboratory in an untidy state. He left an open petri dish, inoculated with *Staphylococcus* bacteria, beside an open window. In Fleming's absence, this dish became contaminated by a comparatively rare strain of airborne mold, *Penicillium notatum*. Spores of the mold probably drifted into the laboratory through the open window from an adjacent laboratory, where researchers were using molds in their research on the development of vaccines. Because the summer had been unseasonably cool, the staphylococci had barely grown on the plate, and the contaminating mold was able to take hold. After warmer weather returned, the bacteria began to multiply and expand across the plate, but they were inhibited in certain places by the presence of the mold. It was these "zones of inhibition" around the mold colonies that

Fleming discovered when he returned from vacation. Recognizing the significance of the mold's ability to inhibit the bacterial growth, he cultured the mold and collected large quantities of the bacteria-inhibiting substance it produced, which he called "mold juice," for further experimentation. Later, after identifying the mold as belonging to the genus *Penicillium*, he renamed the active substance "penicillin." Although Fleming was not the first scientist to have noted the antibacterial properties of *Penicillium*, he was the first to extract and purify penicillin and to search for further applications.

First Tests of Penicillin

Fleming first used penicillin as an experimental tool to aid his research into the development of an influenza vaccine. He observed that it was ineffective against *Haemophilus influenzae*, a bacterium believed (wrongly) at the time to be the cause of influenza. In late 1928, Fleming explored its use in treating infections in blood and human tissue—with limited success. These tests appeared to show that penicillin, at least in its impure form, was only effective when delivered at exceptionally high concentrations, too high for patient treatment. Similarly, Fleming's experiments in 1929 on rabbit organs infected with *Staphylococcus* appeared to show that penicillin would not penetrate beyond the surface of organs. These experiments led to an underestimation of penicillin's value in medicine.

Fleming's early tests had a crucial flaw—they tested efficacy against bacteria in the blood and tissue of dead animals, rather than against infections in living animals. Fleming's experiments on live animals were limited to testing low doses for toxicity. In the late 1930s, a team of researchers, led by Ernst Chain and Howard Florey at Oxford University, investigated how penicillin destroyed the cell walls of bacteria, research that was aided by Fleming's strains of *Penicillium* mold. Chain independently developed a method of extracting penicillin that was similar to that of Fleming, and he reached a number of similar conclusions about the nature of the substance. Crucially, Chain also theorized that the product he had extracted required further purification to increase its potency if it was to be put to therapeutic use. Chain managed to secure three years of funding from the UK Medical Research Council shortly after the outbreak of the Second World War in 1939. This was followed by a five-year, $25,000 grant from the Rockefeller Institute, which had already been funding US-based research into new antimicrobials.[3]

These funding commitments—which were very significant investments for their day—brought Chain together with Norman Heatley, with whom he worked further on the purification process. They drew on a newly developed technique from Sweden, freeze-drying, which enabled them to produce small quantities of a fine brown powder to begin testing. The first of these tests began in May 1940, on mice inoculated with lethal doses of streptococcus bacteria, with startling results. The mice treated with repeated doses of penicillin recovered. The mice that did not receive penicillin, or received only a single dose, died. These first experiments in animals held significant promise, and Chain and Florey published a paper in the journal Lancet just a few months later.

The Oxford team began producing more penicillin, scaling up the process as much as their equipment and facilities allowed, even resorting to improvised production lines using bedpans and milk churns. The limited amounts that Chain's team could produce constrained their ability to begin clinical testing on any scale. Nevertheless, in early 1941 penicillin was tested in humans for the first time, less than a year after being tested in animals for toxicity. The first recipient of penicillin for therapeutic purposes was Albert Alexander, an Oxford police officer who was critically ill with a serious infection and had already lost an eye.

Faced with such an ill patient, Florey was persuaded to test the new drug on him. The results were almost as pronounced as those seen in experiments with mice: over four days of treatment with penicillin, Alexander showed rapid improvement, and the infection went into remission. Tragically, the supplies of penicillin were quickly exhausted and Alexander's condition began to deteriorate again. Less than a month after this relapse, he died.

The brief success of Alexander's treatment provided another important indicator of penicillin's enormous potential, but it also underlined the need to find a way to produce larger quantities of the drug.

Mass Production of Penicillin

Supported by the Rockefeller Institute, Florey visited the United States in 1941 and searched for collaborators. In Washington, DC, he piqued the interest of scientists at the US Department of Agriculture (USDA), who immediately seized upon the potential application of fermentation techniques derived from those used by brewers. Tests began within days at USDA research laboratories, initially using *Penicillium* samples from Oxford. Yields of the mold increased eightfold over the course of the autumn.

The success of the work at the USDA laboratories led to a pivotal meeting in New York City in December 1941. Four US pharmaceutical companies (Pfizer, Merck, Squibb, and Lederle) were viewed as having the ability to apply these new fermentation techniques to the large-scale development of penicillin as a therapeutic agent. Merck immediately agreed to a public-private collaboration; Squibb and Pfizer eventually followed suit. A number of other companies (including a collaborative Midwest group of six major pharmaceutical manufacturers) also began development of penicillin, which was being treated as public property rather than as a patented invention. In addition, the US military collected soil samples from locations around the world, in an attempt to find strains of *Penicillium* that produced higher concentrations of the precious extract. Promising soil samples were found in Cape Town, Mumbai, and Chongqing—but the best sample came from an overripe melon bought in a Peoria fruit market near the USDA's research laboratory. These efforts, as well as collaboration with the UK companies Boots, Glaxo, and Burroughs-Wellcome, enabled penicillin production to escalate over the next two years.

In 1942, US researchers began to experiment with use of penicillin in patients. In March of that year, the first American patient was successfully cured of a serious staphylococcus infection. Later in the year, only eighteen months after Florey's first test on Albert Alexander, penicillin was used in the first large-scale treatment following a catastrophic fire during the 1942 Thanksgiving weekend at the Cocoanut Grove nightclub in Boston. The fire killed more than five hundred revelers and left more than two hundred survivors with severe burns. As doctors struggled to manage *Staphylococcus aureus* infections in these patients, authorities allowed Merck to release a 32-liter batch of liquid penicillin culture, which was rushed under police escort from Merck's facility in New Jersey to Boston. Its use on the burn victims, along with experimental techniques using blood plasma to replace lost fluids, yielded excellent results.

These successes provided dramatic evidence of penicillin's potential. The US and UK governments recognized its possibly enormous contribution to the allied war effort, and pharmaceutical companies seized upon the commercial possibilities. By the end of 1942, output of penicillin had increased 140,000-fold over the amount that Florey's Oxford labs had produced just two years previously. With the US government anxious to build stockpiles for future war efforts, and companies keen to gain competitive advantage, output continued to soar in 1943 and 1944. US pharmaceutical companies produced 400 million units of penicillin in the first five months of 1943, and this number rose to 20 billion for the remainder of the year. At the beginning of 1944, Pfizer manufactured 4 billion units each month. By the end of that year the company had increased production twenty-five-fold to become the world's largest manufacturer of the drug, with a monthly output of 100 billion units. Penicillin was available to allied troops for the D-Day landings of May 1944 and for the remainder of the war, something that undoubtedly saved many lives on the battlefield. In 1945 the US government removed all restrictions on the release of penicillin to the civilian market. The first antibacterial blockbuster drug was born.

The Golden Age of Antibiotic Discovery

After the frenzied efforts to establish large-scale manufacturing facilities for penicillin, the advent of peace in 1945 ushered in a golden age of antibiotic discovery. Pharmaceutical companies sought to exploit both the medical and commercial potential of penicillin and attempted to find new products that would have similarly miraculous properties.

Recognizing how commercially lucrative these new products could be, many of the pharmaceutical companies marketed their products extensively toward the general public. In the United States, consumers enjoyed virtually unfettered access to penicillin until the early 1950s; it was available over-the-counter in neighborhood drugstores. Because of the amount being produced and the competition between companies providing it, the cost was just a few cents per dose.

Even at this early date, there was some concern about the potential dangers of unrestricted access to penicillin, and some countries, including the United Kingdom, placed greater restrictions on access. In his acceptance speech upon being awarded the Nobel Prize for Medicine in 1945, Fleming expressed his concerns about misuse of the drug by laypeople:

> The time may come when penicillin can be bought by anyone in the shops. Then there is the danger that the ignorant man may easily underdose himself and by exposing his microbes to non-lethal quantities of the drug make them resistant. Here is a hypothetical illustration. Mr. X. has a sore throat. He buys some penicillin and gives himself, not enough to kill the streptococci but enough to educate them to resist penicillin. He then infects his wife. Mrs. X gets pneumonia and is treated with penicillin. As the streptococci are now resistant to penicillin the treatment fails. Mrs. X dies. Who is primarily responsible for Mrs. X's death? Why Mr. X, whose negligent use of penicillin changed the nature of the microbe. Moral: If you use penicillin, use enough.

Fleming's admonition about the proper use of penicillin may have gone unheeded in some quarters, but the phenomenon of resistance was not unknown. The sulfonamide class of antimicrobials that had been developed before penicillin had quickly lost effectiveness because of the emergence of resistance. However, the threat posed by rising resistance was largely perceived to have been mitigated in the years that followed, as pharmaceutical researchers in the United States and beyond enjoyed an extraordinarily productive period of antibiotic discovery, ensuring a plentiful supply of new products to the market.

During this period of fierce competition between pharmaceutical companies, significant efforts were invested in antibiotic discovery programs that used systematic screenings of natural products similar to those that had been carried out during the early 1940s—often involving collecting and testing soil samples from far-flung places in an attempt to identify microbes that had natural antibacterial properties. Companies invested in their own in-house research facilities, as well as partnering with public and academic research institutes. However, the collaborative approach seen during the war steadily unraveled as companies sought to eke out competitive advantages where they could.

For instance, Pfizer, the world's largest producer of penicillin in 1945, was stung by the collapse of the drug's price and the rapid arrival of new competing antibiotics such as chloramphenicol (the success of which propelled its patent-holder, Parke, Davis & Company, to the position of the world's largest pharmaceutical company in 1949). To maintain its position within the market, Pfizer invested in a discovery program to obtain and screen nearly one hundred thousand soil samples from around the world. After eighteen months, a promising compound was extracted (again, ironically, from close at hand—a soil sample taken at one of the company's own facilities) that would be patented in 1949 as oxytetracycline. Having invested $4 million in its development, Pfizer then spent twice as much again over the next two years on an enormous campaign to market the new antibiotic directly to consumers—a controversial move that predictably saw the new product capture a sizeable market share by 1951 (equal to that of its rival, chloramphenicol) and re-establish the company's position in the sector.

This episode exemplifies the level of excitement and competitive activity that the arrival of antibiotics sparked within the pharmaceutical industry in the postwar years—a paradigm that came to be the prevailing business model of the sector as a whole in the decades that followed.

Antibiotics as the Foundation of Modern Medicine

Antibiotic discovery revolutionized the medical and pharmaceutical landscape. Penicillin presented an entirely new treatment for previously incurable illnesses, creating excitement within the pharmaceutical industry and in the general public as news of a "wonder drug" spread. Stories about the miraculous properties of penicillin first appeared in the media during 1943 and 1944, ranging from *Reader's Digest* tales of children being saved from severe endocarditis (a deadly infection and inflammation of the heart), to the claim that the actress Marlene Dietrich had been saved from a potentially fatal bout of pneumonia while touring southern Italy entertaining American troops. *Time* magazine placed Alexander Fleming on its cover and suggested that when considered "under the aspect of eternity," penicillin's arrival might have been a more important event than the success of the allied landings in Europe.

The often breathless and hyperbolic media coverage of a supposed miracle cure may have been fanned by the recognition of penicillin's use during the war, but it also reflected genuine excitement within the medical community about the potential for its use in everyday settings. Physicians began to use penicillin for a range of common infections, from sexually transmitted diseases, like syphilis and gonorrhea, to common sore throats. Mortality from pneumonia declined in the years immediately following the war, as did infant mortality. As sulfonamides and subsequently penicillin began to be used during the 1930s and 1940s, infant mortality in the United States fell by more than fifty percent. Patients who might have died in the past were now surviving, thanks to penicillin.

Another important antibiotic discovered in the 1940s was streptomycin. Discovered by Selman Waksman and Albert

Schatz at Rutgers University, with backing from Merck, strep-tomycin was the first antibiotic effective against *Mycobacterium tuberculosis*, the bacterium that causes tuberculosis. As it did with penicillin, the US government collaborated with Merck and other pharmaceutical companies to produce streptomycin at enormous speed. The discovery earned Waksman a Nobel Prize, and although long-term and high-dosage use of streptomycin was later found to cause deafness in some patients, it transformed the treatment of TB.

Antibiotics brought to market in the 1940s and 1950s provide the foundation of our formulary of antibiotics today. Penicillin and its many derivatives account for almost half of all antibiotic prescriptions in England in 2015. And the availability of effective antibiotics has been vital to the development of medical and surgical interventions now considered routine. Solid organ transplant, for instance, was first pioneered in the 1950s, and 120,000 procedures take place globally each year. The growing success of organ transplant surgery is due in no small part to the effectiveness of immunosuppressive drugs that reduce the probability of organ rejection. Yet patients taking these drugs are extremely vulnerable to opportunistic infections, both during their hospital care and in the months after. This, in turn, creates a need for effective antibiotics to treat these infections for months and sometimes years after surgery. Without antibiotics, successful transplant surgery would not have been possible—the risks and consequences of bacterial infections would have been too great.

The transition from the early nineteenth century, when infectious disease was rife and largely untreatable, to an era of treatable infectious disease and increased life expectancy, was profound for many high-income countries, but it cannot be ascribed solely to the discovery of penicillin in 1928 and the subsequent development of other antibiotics. Many other factors were at play, including rising affluence, improved nutrition, and the introduction of improved public health measures such as proper sewage systems and accessibility to clean water. Many countries in the world are still struggling with some of these challenges today. However, antibiotics did change the definition of what was medically possible, and infections that had once been simply untreatable immediately became a far lesser cause for concern. Antibiotics were truly the world's first blockbuster drugs, permanently transforming the pharmaceutical industry and becoming a reliable source of profits in the decades following the Second World War. Demand for them boomed, and they were the basis for many pharmaceutical advances that followed. But this bounty laid the foundation for the practices that have led to the threat of antimicrobial resistance: rising, often excessive, consumption of antibiotics, and a market (in the developed world, at least) that is saturated by a cheap and ready supply of these drugs.

Two

The Rise of Resistance

Derek Butler does not take antibiotics for granted. Butler first realized the dangers of antimicrobial resistance in 2003 when his stepfather, John Crews, was rushed to the hospital following a heart attack. When we interviewed Butler in November 2016, he described what happened: "He'd picked up an infection in the hospital and the hospital had said to us: 'It is just an infection; do not worry, we are treating it.' But nothing seemed to respond, his condition did not get any better—he did not get any worse for a while, but he just did not pick up. When we queried this they said, 'Well we're changing the antibiotics, we're going to have a go with another type,' and they kept changing the antibiotics that they were treating him with." Butler's stepfather had contracted methicillin-resistant *Staphylococcus aureus*, commonly known as MRSA. The antibiotics did not work because the bacteria causing the infection had become resistant to the methicillin and other similar antibiotics. Butler's stepfather passed away. "He died fifteen weeks later—from what they called organ failure, because of his heart attack. What we found out afterwards was that he was profusely infected with MRSA, and the MRSA responded to nothing—not a single antibiotic."

Learning that some infections will not respond to any antibiotic is quite shocking for many people. In his grief, Butler sought answers. How common was antimicrobial resistance? Was it just for this type of infection? Which countries were most affected? He found out that the situation was even worse than he thought.

"It is not just MRSA," said Butler. "There are other bacteria that are building up resistance: *Clostridium difficile, Enterobacter,* and different bacteria like that. I suddenly thought—right now this is becoming not just other people's problem, it is becoming mine as well." Butler cofounded and is now chair of MRSA Action UK, a charity that supports people who have been affected by health-care associated infections and is dedicated to raising awareness of this problem.

Butler is a great believer in the power of grassroots activism and "putting faces to numbers." He explained, "We found from what we've done over the last eight to ten years that it is a mix of people [affected]. It is not just older people, it is younger people, healthy people. We're talking of people from different religions and creeds. I have a saying: 'bacteria know no boundaries'— whether it is a country's boundaries or human boundaries, it will affect every person on this planet. That's the key. People need to understand this is going to affect everyone."

MRSA Action UK and other organizations have a very important role in communicating the problem of antibiotic resis-tance to the general public, especially since Butler and others can tell their personal stories of what it is like to lose someone to this scourge. Although the level of awareness has improved in the United Kingdom over the last ten years, Butler still be-lieves there is a lot more work to do. "When the Department of Health did a study a few years ago . . . a lot of people were not sure what AMR [antimicrobial resistance] was. They thought it was their body building up resistance to the antibiotics, not bacteria. The AIDS campaign was very powerful in the 1980s. Very, very powerful because you had celebrities on there, and celebrities unfortunately dying from AIDS. You had a short campaign explaining the dangers of AIDS and the threat that it posed. . . . This is what we need now. We need to communicate the message and make sure that people understand it."

After losing his stepfather to a resistant infection, Butler faced another crisis in his family that also involved a bacterial infection—but had a different outcome. Butler's father was hospitalized in 2010 with an *E. coli* infection that was successfully treated. "His chances of survival at eight-nine were pretty remote. But the doctor at the time said that he was going to treat him very aggressively with antibiotics—the right antibiotics." The doctor used a diagnostic test that supplied results in only four to six hours. The results of the test enabled the doctor to switch to the correct drugs and possibly save Butler's father's life. "Within twenty-four hours my dad was sitting up in bed having a meal and talking and chatting away to us." This story highlights how miraculous antibiotics can be when they are effective. They have the ability to save someone who is days or even hours away from death. As Butler put it, "They are the most important drugs in the medical arsenal."

How Resistance Forms

Bacteria adapt and survive according to the same rules of evolution that apply to other living things on earth. All known living creatures replicate themselves either by directly copying their genes (asexual reproduction) or by coming together in pairs to create a third organism that has a mix of both parents' genes (sexual reproduction). During the process of reproduction, mistakes occasionally occur while the genetic code is being copied or transferred. These mistakes, called mutations, cause a change in the DNA sequence of a gene. Mutations can also be caused by certain environmental factors, like radiation. When a mutation is present in an organism, it can be passed on to some or all of the organism's offspring, and they can then pass it on again. (In humans and other multicellular organisms, a mutation is passed on only if it is present in the reproductive cells.) Most mutations are either detrimental or benign. If the mutation causes a disadvantage—even a slight disadvantage—then, over time, the descendants who receive this mutated gene will have fewer descendants themselves, and the gene will eventually disappear. Occasionally, however, a mutation is advantageous to the organism. In that case, organisms with that mutation will leave more surviving offspring, and the mutation will become widespread. Bacteria produce asexually, and genetic material in the chromosome is passed on from one bacterium to its daughter cells when it divides. But some of the genes in bacteria reside outside the chromosome in small structures called plasmids, which can be passed from one bacterium to another when they come into contact. Most antibiotic resistance is transmitted by means of plasmids; genes on plasmids can spread much more quickly than those in the chromosome, and they can spread to unrelated types of bacteria.

Picture two individual bacterial cells, one with only a weak defense against an antibiotic, and the other with a better defense because it has a gene mutation for a trait such as an altered cell membrane that keeps antibiotics out. When treated with an antibiotic, the likelihood of survival is greater for the bacterium with the stronger membrane. This drug-resistant bacterium has a selective advantage; its daughter cells will multiply, whereas the sensitive bacterium will die and leave no offspring. A population of bacteria does not develop complete resistance to an antibiotic overnight. Instead, selective pressure leads to greater and greater resistance over time as the antibiotic-resistant bacteria increase in abundance. Eventually, the antibiotic no longer works for an infection caused by that particular strain of bacteria.

Evolution occurs more quickly in microbes than in large animals. It has been 1.9 million years since our ancestors began to walk on two legs. They were only about 1.2 m. (4 ft.) tall, weighed about 32 kg. (70 lb.), had a brain that was about half the size of ours, and would barely be recognizable as human-like. The first modern humans, *Homo sapiens*, are thought to have appeared about two hundred thousand years ago. The rate of evolutionary change is slow for humans because the mean generational interval is about thirty years. There have only been about sixty-three thousand generations between us and the first human ancestors who walked upright. The generational interval for bacteria, by contrast, is very short. *E. coli*—a common bacterium in humans—has a generational interval of about twenty minutes. Every two and a half years, *E. coli* bacteria go through the same number of generations as humans do in two million years; thus, advantageous mutations can spread very quickly.

The vast majority of antibiotics today have not been synthesized from scratch but are based on bactericides (bacteria-killing compounds) that already exist in nature and that researchers have adapted for medical purposes. A good example of this is penicillin, discussed in Chapter 1, which is produced by certain types of fungi to protect themselves from bacteria. The ability to produce penicillin appears to have evolved in these fungi over millions of years. During this time, some bacteria evolved to counter this defense by producing penicillinases, enzymes that degrade penicillin. This natural arms race between fungi and bacteria is not dissimilar to the scenario we now find ourselves in. Although penicillinases and other resistance mechanisms evolved long before Fleming's fortuitous discovery, they are spreading much more rapidly now because of the creation of greater selective pressure resulting from human use of antibiotics.

In a study by Michael Baym and colleagues at Harvard Medical School aimed at understanding the evolution of resistance, researchers created a 1.2-meter-long petri dish with five sections, each with a different level of the antibiotic trimethoprim. The first section contained no antibiotic, and the second contained three times the minimal amount of antibiotic normally needed to prevent bacterial growth (known as the minimum inhibitory concentration, or MIC).[1] The third section contained 30 times MIC; the fourth, 300 times MIC; and the fifth, 3,000 times MIC. Researchers covered the petri dish in a substance that made it easy for the bacteria to migrate, along with a dye that turned them white, and then placed a small number of naturally occurring *E. coli* in the antibiotic-free area. These bacteria had no previous exposure to human antibiotics and thus no natural resistance to them.

From photos that were taken every 44 hours, we can see how quickly the bacteria spread across the petri dish (Figure 2.1). At the start, there are so few bacteria that you cannot see them on the dish. Within 44 hours, the total area of the antibiotic-free section has been used up, and the bacteria need to move elsewhere to find new resources. One small group has just been able to break into the 3 MIC section. Another 44 hours later, and most of the 3 MIC section has been occupied by bacteria, but they have not yet entered the 30 MIC section. At the 132nd hour they have occupied the entire 3 MIC section and have managed to enter the 30 MIC section. The fifth frame of the photo shows the point at which they enter into the 300 MIC area, and in the sixth frame we observe the tentative move into an area that has 3,000 times the concentration of antibiotic normally needed to defeat these bacteria. By the 264th hour, 11 days after the experiment started, some bacteria are thriving in this 3,000 MIC zone.

It is often easier for bacteria to become resistant to an antibiotic if the concentration is increased slowly than if they are treated with a high concentration at the start. In order to calculate the strength of this effect, the authors of this study tried several different combinations of antibiotic concentrations on the petri dish. They showed that when there was a greater increase in concentration from one section of the petri dish to the next, it took the bacteria longer to enter the higher-concentration area.

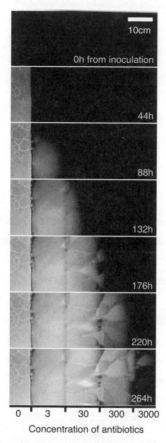

10cm

0h from inoculation

44h

88h

132h

176h

220h

264h

0   3   30   300   3000
Concentration of antibiotics

Fig. 2.1. Bacteria evolve to defeat antibiotics on a petri dish. Each frame, from top to bottom, shows the petri dish at a different number of hours after inoculation. The sections of the petri dish each contain a different level of the antibiotic trimethoprim, measured in units of minimum inhibitory concentration (MIC). By 264 hours (11 days) after inoculation, some bacteria have migrated into the rightmost section and are able to survive in concentrations of antibiotic 3,000 times stronger than would normally be lethal. Credit: M. Baym, T. D. Lieberman, E. D. Kelsic, et al., Spatiotemporal Microbial Evolution on Antibiotic Landscapes, *Science* 353, no. 6304 (2016): 1147–1151.

But the bacteria were eventually able to cross from an area with no antibiotics to one with 3,000 MIC, albeit more slowly. This provides support for the practice of prescribing high doses of antibiotics, particularly new antibiotics, as a way to make it harder for resistant strains to develop. But doing so will just delay, not prevent, the development of resistance.

The Rise and Cost of Resistance

Before penicillin had even been used in people, there was evidence to suggest that resistance might in time stop the drug from being effective. An article published in 1940 noted that some bacteria produced an enzyme that deactivated penicillin. However, the excitement for the "cure" appears to have drowned out those voices that warned of potential problems, including Fleming's. Some people held out hope that the problem of resistance would be minimal. A 1952 publication by Rollo and Williamson presented evidence that penicillin resistance could develop in laboratory mice, but the authors concluded that the increase in resistance was small, and they noted that resistance had not become a major problem for treatment of infections with arsenic-containing drugs: "Syphilis has now been treated with arsenicals for about 40 years without any indications of an increased incidence of arsenic-resistant infections, and this work gives grounds for hoping that the widespread use of penicillin will equally not result in an increasing incidence of infections resistant to penicillin."

By the mid-1950s, a sharp rise in penicillin resistance led to concerns about its continued usefulness. Penicillinase-producing strains of bacteria had become universally present in hospitals. In response, chemists produced new antibiotics, including methicillin, a semisynthetic derivative of penicillin. Once these new drugs were introduced, however, resistance to them developed almost immediately. Indeed, MRSA went on to become one of the most well-known and deadliest types of resistant superbugs in the 1990s and early 2000s.

Carbapenems, a class of antibiotics that were first licensed in 1985, are the "last-line" drugs we currently rely upon to protect us against bacteria that have become resistant to older antibiotics. They are used particularly for infections caused by bacteria in the *Enterobacteriaceae* family, which includes *E. coli*, *Klebsiella*, and *Salmonella*. Carbapenems must be delivered by IV line—meaning that patients need to be in the hospital—which results in a reduction of unnecessary use. However, in recent years there has been a significant rise of resistance against carbapenems. In the United Kingdom, rates of *E. coli* resistance almost tripled between 2008 and 2013, and rates for *Klebsiella pneumoniae* more than doubled in the same period. Between 2008 and 2012, carbapenem-resistant *Enterobacteriaceae* (CRE) rates in the United States increased five-fold, and they rose dramatically across Continental Europe as well, with drug resistant *Klebsiella pneumoniae* found in more than 25 percent of bloodstream infections in Italy, and 50 percent in Greece. In India, 58,000 newborns die of resistant infections every year. On the Indian subcontinent, 200 million people carry carbapenem-resistant *Enterobacteriaceae* in their gut, where they are harmless; if they spread to another part of the body and cause an infection, however, carbapenems will not work against them, making treatment very difficult.

Carbapenem-resistant *Enterobacteriaceae* produce enzymes called carbapenemases that disable carbapenems and, as shown in Table 2.1, very few current drugs are effective against them. Most antibiotics fail to treat these bacteria more than 50 percent of the time. Two antibiotics are effective more than 50 percent of the time but are still unable to treat at least 10 percent of infections. Only one drug, colistin, fails less than 10 percent of the time, the preferred rate for antibiotic treatment. Although colistin can usually kill these bacteria, it has serious side effects: it can cause kidney failure or nerve damage. If a patient goes into septic shock (a quick drop in blood pressure, which can be fatal if not treated promptly), a doctor will not hesitate to prescribe colistin. But it must be said that we have failed in controlling antimicrobial resistance whenever a physician has to make a choice be-

Table 2.1 Drug resistance rate of selected antibiotics against carbapenemase-producing *Enterobacteriaceae*

Antibiotic	Drug resistance rate (%)
Effective	
colistin*	3.3
Often fails	
tigecycline	23.2
amikacin	31.0
Usually fails	
gentamicin	58.0
ciprofloxacin	66.3
tobramycin	67.7
meropenem	74.2
aztreonam	82.5
imipenem	82.8
ceftazidime	88.8
cefotaxime	95.5
ertapenem	98.2
pipericillin-tazobactam	99.3
co-amoxiclav	99.7
piperacillin	99.7
ampicillin	100

* can cause kidney damage

Note: Proportions for metallo- and nonmetallo-carbapenemase-producing *E. coli*, *Klebsiella*, and *Enterobacter/Citrobacter* have been averaged for simplicity.

Source: Antimicrobial Resistance and Healthcare Associated Infections (AMRHAI) Reference Unit of Public Health England.

tween a safe drug that might not work and a dangerous but effective one. This kind of difficult choice is becoming all too common.

Quantifying the Problem

Many studies have sought to estimate the scale of the problem of antimicrobial resistance in specific localities; however, the United Kingdom's Review on Antimicrobial Resistance was the first study that attempted to estimate the global burden of this problem. In its 2014 report, the Review calculated, as a low estimate, that more than 700,000 people die every year worldwide as a result of antimicrobial-resistant infections. This amounts to about one in every eighty deaths. Reliable data were available only for the United States (from the US Centers for Disease Control and Prevention [CDC]) and the European Union (from the European Centre for Disease Prevention and Control [ECDC]); for other parts of the world, the authors assumed that the rates of death due to drug resistance were the same as the current death rate from antimicrobial resistance in the United States. This total is useful for talking about the magnitude of drug-resistant deaths worldwide. However, we know that resistance rates are much higher in low- and middle-income countries than in wealthier countries because of poorer sanitation systems, lower-quality health-care systems, and weaker controls on the dispensing of antibiotics.

Since this low estimate was released in 2014, more data have become available. A 2016 study led by Cherry Lim estimated that 19,122 people die in Thailand every year as a result of drug-resistant infections, or about 28.6 per 100,000 people, about four times the rate believed to exist in the United States. Lim and her colleagues painted a bleak picture of the situation in Thailand, where an increasingly large number of infections can only be treated with highly toxic drugs. Survival rates are low, and the health-care system is struggling. The researchers criticized the lack of global surveillance and presented a low-cost way of estimating deaths from antimicrobial resistance, using microbiology laboratory and hospital databases of nine public hospitals in northeast Thailand from 2004 to 2010. This approach provides a promising and efficient way to track the effect of antimicrobial resistance worldwide. The authors concluded that "the prevalence and mortality attributable to multi-drug resistance in Thailand are high. This is likely to reflect the situation in other low- and middle-income countries."

Based on this study, we have tried to calculate an estimate of deaths caused by antimicrobial resistance that reflects the much higher fatality rate in low- and middle-income countries. We assumed, as a baseline, that all middle-income Asian countries had a drug-resistance mortality rate similar to that of Thailand. Although most non-Asian, non-European countries and low income Asian countries probably face far greater resistance problems, because we lack accurate data, we used US mortality rates for these countries. We also used US mortality rates for high-income Asian countries, and we used the ECDC figures for countries in the European Union.

With this new data, we calculate that the total number of people dying every year from antimicrobial resistance is approximately 1.5 million. This number of deaths is greater than the number of people who die worldwide from road accidents (1.2 million), and the same as the number of people who die from diabetes.

Under an initiative known as the Fleming fund, the UK Department of Health is funding the building of infectious disease surveillance labs across the world. In the upcoming years we should have much better data on the burden of resistance in different locations. The Institute for Health Metrics and Evaluation (based at the University of Washington) is also being commissioned by the UK Government and the Wellcome Trust to analyze the global burden of antimicrobial resistance and to estimate mortality rates. These initiatives are years away from producing results. Nevertheless, existing estimates clearly show that huge numbers of people are dying as a result of antimicrobial-resistant infections every year. We should always try to improve predictions, but we must take action now.

The Economic Cost of Resistance

In addition to the cost in human lives, antimicrobial resistance places a burden on health-care systems and resources. People with resistant infections spend more time in hospitals, require greater supervision from doctors and nurses, need more expensive drugs, and often have to be isolated from other patients. In the United States, for example, it costs an average of $16,000 to treat a patient with methicillin-susceptible *Staphylococcus aureus*, and there is an 11.5 percent chance she will die. If she has MRSA, the resistant variant, the cost is $35,000, and her chance of dying is 24 percent.

We calculated the economic cost of this problem using three methods that are common in economics. The first is to assess the additional health-care costs of a disease and its treatment. The second is to look at lost productivity, including lost time at work and cancellation of work travel. The third involves calculating the social cost of illness or death using what is called the value of a statistical life—so named because it indicates how much we will pay to reduce risk to such an extent that one less person is expected to die. (This value might not be the same as that placed on preventing the death of a particular individual who is in harm's way.) This last method of evaluation aids governments in deciding how to efficiently spend resources to prevent deaths. For example, installing streetlights is expensive but reduces mortality from accidents. Imagine two extreme scenarios. In the first it is estimated that installing streetlights would prevent fifty deaths over the next fifty years, and would cost $50,000 for installation, electricity, and maintenance during this period. At a price of only $1,000 per expected life, the decision to add streetlights seems fairly obvious. But what if the cost were $1 billion, and only one death would be prevented? In that case, the government could probably prevent significantly more deaths by investing the money elsewhere. The question of where to draw the line between investments to save lives is rarely this clear-cut, and the value of a statistical life can be a useful concept for making such decisions when the answer is unclear.

Unfortunately, there is no good way to determine what the value of a statistical life should be. In November 2016, we asked Nicholas Stern, chair of the Grantham Research Institute on Climate Change and the Environment, about how to determine this value. He told us, "The value of a statistical life is often for these purposes around 100 times GDP [gross domestic product] per capita. That number emerges from studies of what people are willing to spend on road improvements that reduce mortality, or hospital improvements and so on. The number of 100 times GDP per capita emerges from watching government behavior and asking what they are willing to spend to save a life. In a number of countries, such as the USA and the UK, such a number is used explicitly in terms of 'dollars per life.' It is a useful number, even though it should not be taken too literally; it must be used with judgment." Stern went on to say that it is neither possible nor crucial to determine these figures precisely. They are, however, very useful for giving a feel for the scale of the problem.

As an example, in 2007, the UK Department for Transport was willing to spend $3 million to prevent each additional expected fatality when designing roads or adding safety features such as street lighting. The UK's GDP per capita was $48,300 in 2007, so that means the government was willing to pay 62.5 times the GDP per capita to prevent a death. In comparison, the US Department of Transportation was willing to spend $5.8 million to prevent an expected fatality in 2008. The US GDP per capita was $48,400 per year, which comes out to a willingness to spend 119.8 times the GDP per capita to prevent a fatality.

It is difficult to calculate the impact of the health resources currently spent trying to treat drug-resistant infections, or the cost of lost productivity, when we do not have accurate data on how many people have died or will die from these infections. Given the difficulty in calculating these figures, it is not surprising that existing estimates are both sparse and inconsistent. Consider these two disparate estimates. A study by the European Medicines Agency on the cost of antimicrobial resistance in the European Union estimated the amount spent by all of the European Union's health-care systems to be €900 million ($1.062 billion), or about 0.06 percent of total health expenditures and €600 million in productivity. A study by the CDC, however, came up with much higher numbers; it estimated that health-care costs associated with drug resistance are about $20 billion per year in the United States, while productivity costs amount to an additional $35 billion. Drug resistance reduced US productivity by about 0.23 percent. There are more drug-resistant infections in Europe than in the United States, so it is hard to understand why costs in the United States would be 20 times more than in Europe, and the percentage of health-care expenditures would be 12.5 times more. Because both studies used methodological approaches that are more likely to undercount than to overcount the number of people with resistant infections, we have chosen to use the US figures, which we believe to be more accurate, for estimates of global cost. Using the CDC estimates of cost and lost productivity and applying them to other countries, we would expect the total cost of antimicrobial resistance on world health systems to be about $57 billion, and the reduction in world productivity to be valued at $174 billion per year. The latter figure is equivalent to everyone in the world taking an additional half day off work every year.

If, instead of these estimates based on health-care costs and lost productivity, we use estimates based on the value of a statistical life, using the figure of 100 times GDP, the total cost increases to an estimated $864 billion per year. This number could be somewhat inflated, since many of those killed by resistant infections are very old or very ill, and their life expectancies may on average be reduced by only a few years. Nevertheless, these estimates give some indication of the scale of the financial costs.

No matter how you estimate it, the cost in medical resources and lost productivity comes out to many billions of dollars. Most important from a policy perspective, this cost greatly exceeds the cost of investing in solutions, as we discuss in Part II of this book.

Indirect Deaths and Costs

Modern medicine was built upon reliable antibiotics, and health care would look very different in a world without antibiotics. Surgery, cancer treatments, and a vast number of other treatments would be negatively affected, but exactly how is unclear. For this reason, we are not going to attempt to quantify these indirect costs; suffice it to say that they would be substantial.

Another indirect cost that is hard to quantify is the "aversion cost" that accompanies all major infectious disease outbreaks. This refers to changes in behavior resulting from the perceived threat of disease. For example, people would be less likely to travel if they knew that antibiotics did not work, because they would not want to risk getting an infection that could not be treated. The impact on trade and tourism would have a damaging effect in some parts of the world, especially those with less well developed health-care systems or poorer sanitation systems.

When we interviewed him in October 2016, Peter Sands, chair of the Commission on a Global Health Risk Framework and former CEO of Standard Chartered Bank, discussed this issue in relation to the 2015 Zika virus outbreak:

> There is an odd dynamic . . . in the world of infectious diseases. We seem to be getting better at containing the mortality impact, that is, the loss of human lives, because even if we do not know how to stop the pathogen, we respond very quickly. You see this with things like Zika, where, although it is terrible for the individuals involved, actually the number of human lives lost . . . is relatively small in the scheme of things. However, we seem to be getting even more vulnerable to the economic impact, because with social media and TV, travel and the interdependence of supply chains, there are greater spreading effects. There have not been very many microcephaly cases [a brain disorder associated with Zika] in the Caribbean, but that has not stopped the region's honeymoon industry from being devastated. I could see very similar dynamics happening in the . . . world [of antimicrobial resistance]. If you had a very serious resistant outbreak taking place somewhere, the economic consequences of that could spiral very wide and very fast. The conclusion we [the commission] came to was that alongside the issue of disease contagion, you have the challenge of the contagion of fear. And the economic impact is mainly driven by the fear, rather than the mortality, at least in the short term.

Even small actions taken to avoid or treat an illness can have a significant impact on the economy at large and, by extension, human health. The economy is so complex, and the impact of antimicrobial resistance so vast, that it is difficult to estimate how an outbreak will affect fear levels and subsequent behavior, but it is safe to assume that this would also be very large.

The Future of Resistance

What is terrifying about antimicrobial resistance is that, unlike the situation with most diseases, the treatments we have today are likely to be less effective in the future. The burden of disease will rise. We do not know how many new drugs will come to market, and we do not know how diagnostics and other advances could change our ability to treat patients with infections. But in order to understand and prepare for the possibility of a world without antibiotics, we need to make some predictions about what is to come if nothing is done.

To better understand how the future of antimicrobial resistance can be modeled, we asked epidemiologist Marc Lipsitch, of Harvard's T. H. Chan School of Public Health, why it is so difficult to predict changes in resistance. "We do not understand at the individual level how antibiotic treatment drives antibiotic resistance," he said, in a November 2016 interview. "We understand it qualitatively to some extent, but we really do not understand it quantitatively. For example, the population-wide trends very strongly suggest that the more antibiotics are used, the higher the prevalence of resistance. When you compare across countries or across regions within a country, it is a pretty consistent finding. Similarly, the correlation between resistance in one bug [strain of bacteria] and resistance in another, by geography, suggests that something is a natural driver, and the obvious guess would be antibiotic use. But when you get to the individual patient level our understanding gets foggier." Lipsitch went on to explain that most bacteria get their resistant genes from their parent cell or from a plasmid, rather than from a new mutation.

At an aggregate level we understand this, but distinguishing between new and already existing mutations is nearly impossible to do outside of a lab. We do not know, said Lipsitch, why penicillin use causes "penicillin resistance to become more common in the population, and what the exact steps are between the individuals who are getting penicillin that lead to this."

Earlier in this chapter we discussed estimates by the Review on Antimicrobial Resistance of the number of fatalities currently caused by drug-resistant infections. The Review also estimated how much this problem could increase by 2050 if no action were taken. In 2016, the World Bank released its own forecast of the dangers of resistance. Both of these estimates were intentionally broad and intended not to be definitive forecasts, but rather to assist policymakers and the public to understand the scale of the problem.

The best prediction models are simple ones that are easy to understand and critique. To create its models, the Review commissioned the consulting firm KPMG and the RAND Corporation to predict what would happen if the global rate of resistant infections rose by 40 percent for infections with *E. coli*, *Klebsiella pneumoniae*, *Staphylococcus aureus*, the TB bacillus, the HIV virus, and the malaria parasite. This rate was chosen because it matches levels that are already common in much of Asia and southern Europe, and it is plausible that rates could rise to that level worldwide.[2] The models also had to make assumptions about infection rates, that is, how likely it is for an infection to be transmitted from one person to another. KPMG and RAND modeled two scenarios: one assuming current infection rates, and the second assuming twice the current rate. The latter scenario is considered more likely, since people would carry resistant infections for a longer time (because they are harder to treat).

These models predicted that if the rate of resistant infections rose by 40 percent and infection rates doubled, approximately 7.5 million people could die every year from bacterial infections—a five- to ten-fold increase over our estimate of the burden today. (If HIV and malaria are included, this number rises to 10 million.) But these models were based on the assumption that the drug-resistant infections MDR-TB, MRSA, and cephalosporin-resistant *E. coli* or *K. pneumoniae* would be treated the same as they are today, an assumption that is questionable. At the present time, when drugs do not work we have alternatives. For MRSA, we use vancomycin. However, vancomycin resistance itself is rising, and at the time of writing we do not have a good backup for this drug.[3] When cephalosporins do not work, we can use carbapenems or colistin (which can cause kidney failure but almost always kills the bacteria). In the three years since these models were completed, however, it is becoming clear that the scale of carbapenem resistance is worse than previously estimated. In 2015 a new type of resistance against colistin was found in China, and it is now known to be in most European countries as well.

Resistance rates can increase very quickly, and drug development is a slow process. In the best-case scenario, it can take ten to fifteen years from initial investment until a new drug is perfected, tested, approved, and reaches the market. During this time, if resistance rates increased as fast as the Review forecast they might, tens of millions of people would die. We need to develop ways to preempt resistance rather than let it spiral out of control.

Economic Impact of a Rise in Antimicrobial Resistance

Two large studies have calculated the long-term economic impact of drug-resistant infections—one by the World Bank, and the other by the Review on Antimicrobial Resistance. Although the two organizations used different methodological approaches, and a number of uncertainties make forecasting difficult, they found similar results.

To estimate economic burden the Review focused on productivity costs, taking a very narrow view of such costs based solely on the impact of an individual leaving the labor force if they are hospitalized or die of a drug-resistant infection. This was included in a total factor productivity model, a model used to predict future economic growth. The accounting firm KPMG calculated an $82 trillion loss over the thirty-five-year period from 2015 to 2050 ($100 trillion if HIV and malaria are included). This is equivalent to the entire economic output of the United Kingdom over the same period. Their estimate, used by the Review, was that the world would produce 3 percent less per year by 2050 because of drug resistance.

The World Bank, whose reports are often seen as the gold standard in global economic evaluation, released its forecasts of the impact of antimicrobial resistance two years later. The report modeled an optimistic, or "low AMR" scenario forecasting a lesser impact of drug-resistant infections, and a pessimistic, or "high AMR" scenario of greater impact (as well as a third scenario of no change). The reduction in global economic output by the year 2050 was estimated at between 1.1 and 3.8 percent in the respective scenarios, equivalent to between $2.4 trillion and $6.9 trillion per year.[4] The high AMR scenario would cause an economic setback significantly greater than the global financial crisis of 2008–2009. Meanwhile, there would be costs of between $340 billion and $1.3 trillion per year in additional health-care expenses.

Although the low AMR scenario of a 1.1 percent reduction in world productivity might sound small, it is still significant. Indeed, even in this "optimistic" scenario, the World Bank estimated that 8 million more people would live in poverty by 2050. The World Bank, like the Review on Antimicrobial Resistance, estimated that poorer countries would suffer more from drug resis-

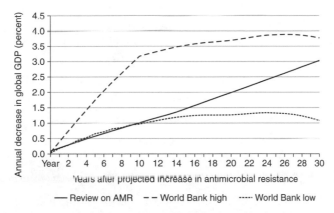

Fig. 2.2. Estimates of annual decrease in global GDP as a result of an increase in rates of antimicrobial resistance. Shown on the graph are estimates by the Review on Antimicrobial Resistance (AMR) and the World Bank (high and low scenarios). Sources: World Bank (2016) and Review on Antimicrobial Resistance (2014).

tance. In the high AMR scenario, the estimate for additional people living in poverty rises to 24 million.

In Figure 2.2, we present a comparison of estimates by the World Bank and by the Review on Antimicrobial Resistance of the economic damage that could result from an increase in drug-resistant infections. Both groups calculated how much additional economic damage would occur on top of the amount already occurring in a particular base year. Because the Review's work was undertaken several years before that of the World Bank, the two groups used different base years. To make comparison easier, we have reformatted the figures so that they are based on the number of years after the projected rise in resistance rather than on a particular base year.

Both estimates show significant decreases in global GDP because of drug resistance. While there is a wide range in the estimates of how antimicrobial resistance will impact the economy, and the World Bank's estimate shows a quicker rise with more of a leveling off than the Review's estimate does, every scenario predicts a major impact on the world economy. What is important for policymakers to recognize is that it will be much cheaper to deal with antibiotic resistance now than after it becomes even more deadly. The World Bank report states, "The risk of AMR further bolsters the economic case for effective and early control of infectious diseases at their source. As with preventing and fighting fires, reducing risks at their source is invariably more effective and more efficient than a reactive stance of waiting for a crisis to develop before responding."

Overview of Potential Costs

We can only estimate the scope of the problem of drug resistance today, and it is impossible to know exactly what it will look like in the future. However, it is clear that if no action is taken, the medical and economic impacts will be significant. Governments might not want to invest in solutions, but they will ultimately pay either way. Any money not spent now will result in substantial costs in the future—not to mention many lost lives. Serious damage to economic productivity (which by extension threatens governments' tax incomes) coupled with the higher costs of health care (which is largely government funded) should provide the impetus to deal with this crisis now.

Highlighting the cost of drug-resistant infections has played an important role in encouraging policymakers, particularly those in government finance departments, to engage with this problem. As former UK chancellor of the exchequer George Osborne told us, "Initially, they [ministers for finance] didn't understand fully why it was an issue for finance ministers as opposed to just health ministers. That remains a problem today, but there is a growing recognition of the financial costs of failing to tackle antimicrobial resistance and the need for financial expertise in developing the solutions."

Antimicrobial resistance has the potential to greatly undermine our health and economic systems. Much of the great medical progress we have seen in treating infections will be undone if we are not careful to protect the advances made by Alexander Fleming and his successors. Not only will infections kill more people, but surgery, chemotherapy, and many other aspects of medicine that are standard today will become much more difficult. Grave economic consequences will be much more costly to governments and society than addressing this problem head-on. Why has so little been done? In Chapter 3, we look at the scientific, economic, and political failures that have led to this problem going unheeded.

Three

Failures in Tackling Drug-Resistant Infections

Helen Boucher of the Tufts Medical Center in Boston talked with us in January 2017 about her experiences as a doctor working in a high-income country. Although her patients have a low risk of contracting most infectious diseases, she has seen an increase in drug-resistant bacteria. "We never thought we'd be in a position [25 years ago] where we'd have to send people with infections home or to hospice [because we could no longer treat them]. . . . We as doctors, as parents, as citizens are alarmed that it has come to this." The problem of drug resistance has existed for a long time, she noted, but "it was not really until the last ten years that we came up against the really complicated scenarios that we're facing now."

Boucher treats two different categories of patients—outpatients and inpatients. Drug resistance can be a problem for both groups. "Outpatients acquire infections from organisms that are resistant to multiple antibiotics. The common story is someone who develops an infection in their urinary system. This includes men with a prostate infection, women with a urinary tract infection. . . . In many cases these are people working and functioning in society who we've had to admit to the hospital and treat with intravenous antibiotics. That problem continues to grow. . . . We're talking about people who should be working, going to school, parenting their kids . . . and we have to put them into an acute care hospital bed to treat them. . . . Most of them we manage well, but occasionally bad things happen and we end up running out of options."

A colleague of Boucher's had a patient who ended up having to go to a hospice after many rounds of failed treatments. Boucher treated another patient "who underwent a number of operations to try and help 'cut out' the infected areas, if you will, and finally we ran out of options for him and he ended up passing away."

"Those that are more critically ill to begin with are in the hospital having procedures and surgeries and they wind up with a drug-resistant infection," said Boucher. She talked about instances of patients awaiting a transplant, "just a few, luckily," where we "had to stop and tell the patient that we have nothing left to offer but hospice care. We've lost a few of those patients and they've never received the transplant they needed."

She believes the biggest problem is giving people antibiotics when they do not need them. This is a problem in all specialties, with patients of all ages, but most often in the pediatrician's office. "If we reach the point where we cannot support someone through her joint replacement, her surgery, or her bone marrow transplant, then we're limiting medical progress, we're limiting medical care. . . . And that has happened. We have been in the position where we have had to say that someone is not a candidate for a transplant, for example, because we could not control their infection." This is a sobering point. Even though it is rare to have to refrain from performing a transplant or a surgical procedure because of an infection, this is not the kind of thing we expect to see at all in a highly advanced health-care system.

Boucher shared her perspective of what it is like to be a physician treating someone suffering from a drug-resistant infection. "So I treat a patient with a urinary tract infection, . . . and I give her a prescription for something oral and send her on her way. Then I get a report two days later that says that the organism that's causing the infection is resistant to the pill I gave her, and other pills. There is no other oral medication. I call the patient to see how she's doing and she's doing terrible. She's having fevers, back pain, she's miserable. She's annoyed with me because what I gave her did not work, and in that context I have to say, 'I'm sorry to say that I have to bring you into the hospital now so I can put an IV into your arm and give you IV antibiotics.' The patient cannot go to work, she cannot take care of her child, she cannot live her life until we get this treated. And that's not that uncommon. . . . What should have been a doctor's visit becomes a two, three, four week problem for this individual."

Boucher went on to talk about what can happen in more serious cases:

Then there's the patient in the hospital who's really sick, who came in for something and developed pneumonia, for example. We're consulted, and we recommend antibiotics based on guidelines and the best available evidence, and the patient does not get better. One to two days later we learn that the bacteria that are growing in their lungs are resistant to what we gave them. And we have to give them something stronger, and perhaps more toxic—to their kidneys, to their bone marrow, to some other system in their body. This is an individual who's already quite sick, perhaps on a ventilator in the ICU. So we have to sit down with their family and tell them that unfortunately the first antibiotics were not effective and we have to go to the next. And sometimes that scenario continues and we continue to get news that is not great, or we start that patient on what is appropriate therapy based on the microbiology that we have on the culture results, and they get better for a few days and then they get worse again. And we send off another sample to the lab, and now the bacteria are resistant to the second antibiotic, and we have to dig deeper. Now the choices are definitely toxic to the kidneys, and we have to sit down with the family and say, "Unfortunately now we have to go to something that we know will hurt your dad, your brother, your husband's kidneys and they might even have to go on dialysis." So this person who walked in might have to go on dialysis, if they live.

Fortunately, this kind of scenario does not occur frequently in Boucher's practice, but it is a common experience in countries with higher rates of drug-resistant infections. As Boucher's stories make clear, drug resistance can significantly disrupt patients' lives even if they do eventually make a full recovery.

How did we get to this situation? A combination of difficult science, failed economic models, academic disengagement, and political short-termism are responsible.

Difficult Science

We interviewed John Rex, a leading scientist who has worked in the area of drug development for the past twenty years, in November 2016. Asked about the difficulty of designing drugs to treat infections, he explained:

It is easy to kill bacteria. It is easy to kill fungi. Steam, fire, bleach, they all work great. But the trick of separating toxicity to humans from toxicity to these microbes is hard. There are only two classes of drugs where the purpose of the drug is to kill something that is alive: one is antimicrobials, and the other is anti-cancer drugs. There are a lot of similarities between these two areas. Living organisms do not want to die. They have many, many defenses designed to enable them to live in different environments, so the fact that you can do this at all is remarkable.

Toxicity is a persistent problem for antibiotic development, often because the mechanism that an antibiotic uses to kill bacteria also damages human cells, particularly in the kidneys or liver. Similar toxicity problems arise in chemotherapy, but the comparatively slow growth of cancer cells makes it easier to monitor their impact.

As the challenges of antibiotic development grow, so do the costs. Margaret Chan, who was then head of the World Health Organization, told us, "All of the 'easy' antibiotics have already been discovered. Drug discovery now is far more complex and costly. Antibiotics are also less lucrative than drugs developed to treat chronic conditions."

One challenge in antibiotic development is that almost all of our antibiotics are based on natural compounds, like penicillin, that have been chemically modified. It is possible that there are few novel antibacterial agents left to discover. We need to think of our current antibiotics as nonrenewable natural resources. Long before we discovered the environmental damage caused by burning hydrocarbons, we were keenly aware that one day the world would run out of coal and oil and that not only should we not waste them, but we should develop renewable resources. Both government and industry plan for the exhaustion of rare earth metals that are needed in electronics and elsewhere. This is not to say that we will never find any new antibacterial compounds; we have not done enough searching over the past few decades to know how many useful new treatments could be out there. However as it is unclear how many more drugs can be found in the future, we should work hard to protect the ones we have, as well as new ones that we find. As we discuss in later chapters, we can conserve the currently available antibiotics by preventing infections in the first place, by reducing unnecessary prescriptions, and by removing antibiotics from the environment.

A second challenge for scientists has to do with how antibiotics work. All current antibiotics are based on three broad mechanisms of action: (1) breaking down the bacterial cell wall or membrane; (2) preventing or slowing the cell's ability to synthesize proteins (which are used to build and repair the cell); or (3) preventing the synthesis of the cell's DNA (which contains the genetic material) or RNA (which directs protein synthesis). To protect themselves, bacteria develop enzymes that disable the antibiotic in question by generating pumps to remove the antibiotic from the cell, altering the cell walls to block the antibiotic's entrance, or making small changes in structure to protect their DNA and RNA. Once bacteria have developed resistance using a particular mechanism of action, new antibiotics using the same mechanism will not work as well, making the entire process of finding antibiotics that work against the resistant bacteria more difficult.

We do not yet know whether these challenges can be overcome or if we face insurmountable obstacles in the development of new antibiotics. But we will not know what is scientifically doable until we invest adequate resources into finding out. In the rest of this chapter, we focus on the economic and political problems that have kept the level of investment too low.

Failed Economic Model

In high-income countries, if a particular disease affects many people, companies usually invest significant amounts of money and time in developing drugs for that disease. This has prompted some people to ask why the current system, where demand for new drugs drives up price, thus increasing supply, has not been sufficient to generate the new antibiotics that we need. In this section we try to answer this question by outlining the reasons why the current economic system is not providing adequate incentives to stimulate development of new products.

Supply Doesn't Meet Demand

One of the first things economics students learn is that when a good or service is scarce, or demand is high, the price rises, reducing the number of people who want that good or service. This also creates an incentive for people to go out and make or find more of it. While this system is far from perfect, and many economists have gone to great lengths to uncover "market failures"—areas where the market does not work properly—almost all economists agree that in most situations, supply and demand is the best system for distributing goods and services.

To illustrate how supply and demand works, consider the oil crisis of the 1970s, when the Arab-Israeli war led to reduced oil production and increased prices. In efforts to protect their citizens from higher costs of fuel, many governments, including the Carter administration in the United States, set a government price for gasoline and imposed a rationing system. This resulted in huge inefficiencies and a bureaucratic mess—people sometimes had to wait for hours in long lines to get their fuel allowance. Some people attempted to cheat the system to secure more fuel than they were allotted. The fuel's price was lower than its value to consumers, so there was no immediate incentive to increase supply. Eventually it was decided that the best way to overcome this problem was to let prices rise to the point where demand and supply met. In the short term, prices increased significantly and queues became shorter; although this was a difficult trade-off, it saved drivers long waits and pushed up productivity. In the longer term, higher prices encouraged oil firms to explore for oil in previously prohibitively expensive areas, such as the North Sea and the Gulf of Mexico. The supply of oil increased and created a new price equilibrium. In short, the price increase encouraged investment and innovation to meet demand.

While a process like the one just outlined does work to increase supply and limit the demand for existing goods and services, it is less effective in encouraging efforts to develop new products. From indoor plumbing and the lightbulb to smartphones and antibiotics, material gains over the past two centuries have been driven by innovation requiring financial investment and risk. In the past, economic systems were poorly set up to reward innovation; it is often cheaper and easier to copy someone else's invention than to create a new one. When new products were copied, those who undertook the original risk gained little reward or return on investment. This resulted in companies coming up with fewer original ideas or investing in areas that were harder to copy. In response, governments created new property rights, most notably the patent system, which encouraged innovation by giving the inventor a monopoly over her invention for a defined number of years and stalling competition. Patents encourage investments in new areas, yet they can also stifle innovation because granting a monopoly can halt further investment in an area. Patents are thus still a far from perfect system.

In the pharmaceutical sector, it normally takes ten to fifteen years to bring a new drug to market, in a process that costs more than a billion dollars. Intellectual property rights then give the company a monopoly over its product for a period of approximately twenty years, depending on the country, before low-cost generic manufacturers can sell the product at a reduced price. Much of this twenty years is spent testing the drug in clinical trials, meaning that companies normally only get about ten years of sales to recoup their investment costs. The original innovator loses almost all of the drug's value at the end of the patent-protection period. If you were to find a cure for cancer or dementia tomorrow, most consumers suffering from one of these conditions would want to purchase it, resulting in high sales revenues. In most areas of medicine, scenarios like this take place when a useful new drug comes to market. Most societally useful medications have a good return on investment over the drug's patented life.

Antibiotics are different. If an excellent new antibiotic is effective against infections caused by drug-resistant bacteria, most public health officials would want to protect it for use in the most extreme circumstances and would discourage it from being sold worldwide. To get the maximum benefit from the drug and prevent the development of resistance, it is important that people not use it frequently. When asked what she would do with a useful new antibiotic, the chief medical officer for England, Sally Davies, said that the drug "would need a stewardship program"—that is, that systems would have to be in place to make sure that the antibiotic was only prescribed when absolutely necessary. Indeed, limiting unnecessary use is essential to keep bacteria from becoming resistant to new antibiotics, and thus essential for our continued health. However, this also means that when a really useful new antibiotic is found, the company that invests in it cannot rely on high sales for return on investment, in contrast to the situation in other areas of medicine, where drugs with high societal need have very high sales rates. In the case of antibiotics, societal need does not correlate with product sales.

Low Prices for Antibiotics

Not only is the volume of sales for a patented antibiotic low, but the price point is lower than for other kinds of drugs. The last seven antibiotics approved by the FDA cost an average of $3,749 per course.[1] Doctors will normally only prescribe these drugs when first-line drugs do not work. These drugs are curative; that is, if the drug treats the patient successfully, the patient will return to full health. In contrast, eleven of the twelve most recent oncology drugs approved by the FDA each cost more than $100,000 per year. Rather than cure the patient, these drugs extend life expectancy. For the four oncology drugs analyzed by researchers Tito Fojo and Christine Grady, patients paid an average of $47,000 for every extra month of life they gained from the treatment. The new hepatitis C drug Sovaldi (sofosbuvir), although it is curative, costs $85,000 per treatment. These drugs have the advantage of large patient bases, unlike new antibiotics.

Drugs that have a low sales volume because they treat rare conditions are known as "orphan drugs." Even though they do not sell as well as most drugs, orphan drugs often have sales levels that exceed those of new antibiotics. Because the market for these drugs is small, the cost is high. Drugs for hemophilia, for example, normally cost more than $200,000 per year, for the lifetime of the patient. Enzyme-replacement therapy for Gaucher disease, a genetic disorder, costs more than $300,000 per year, but only about ten thousand people in the world are known to have the condition.

You do not need an economist to tell you that when price and volume of sales are both low, investment in a product has little appeal. No wonder drug companies do not want to invest in new antibiotics. Before we examine how to rectify this situation, it is important to understand why the prices of antibiotics are so low relative to their benefit and to the market value of other drugs. Several phenomena can help explain these low prices. We look in detail at four factors: the availability of substitutes, the lack of incentives to pay for social benefits (the externality problem), poor patient information, and the free-rider problem.

Substitutes

In economic terms, substitute goods are different goods that are bought for the same purpose and can be interchanged, such as two brands of lemonade. When the price of one of these goods goes down, people buy more of it; demand for the other good falls, and its price then goes down, too. The opposite happens when prices go up. In other words, people switch from one product to the other solely on the basis of price. If two goods are not completely interchangeable, however, people may not switch unless there is a large price discrepancy. For example, in a society where people tend to prefer cola to lemonade, the main cola producer is likely to charge more than the main lemonade producer. However, if the lemonade producer halves the price of the drink, then some cola consumers will probably switch to lemonade. Even though lemonade is the less preferred product, the lemonade producer is still able to limit the price the cola maker can charge.

The same principles apply in the case of antibiotics. As long as there are not problems with toxicity or drug resistance, patients and doctors will happily substitute one antibiotic for another. This leads to patients generally being treated with cheaper, generic drugs rather than newer, more expensive ones that are still under patent protection.[2] Substitutes not only reduce the sales volume of new drugs, they also drive down the price of antibiotics. If a patient has an infection that is resistant to off-patent drugs but susceptible to a new one, the physician would treat the patient with the on-patent drug if the price were similar. However, if the on-patent drug costs hundreds of times more than the older drug, the physician might decide to use the older one, hoping to cure the patient by using a higher than normal dose.

In part because of the availability of cheap substitutes, only 12 percent of the revenue from antibiotic sales is for patented antibiotics. The total global market for antibiotics is $40 billion per year, of which patented drugs make up only $4.7 billion. In 2015, there were fifteen pharmaceutical drugs that each generated more revenue than all patented antibiotics combined, including the top-selling arthritis drug Humira, which had sales of $14 billion, and the hepatitis C drug Sovaldi, at $13.86 billion.[3] Since generic antibiotic sales do not generate any incentives for companies to undertake research and development, the low revenue figure for patented antibiotics is worrying.

Externalities

An action is said to have an external effect, or externality, when it affects a third party that does not have a say in the decision. For instance, when you smoke a cigarette in public, you choose to pay some cost (both financial and health) in exchange for the pleasure you gain from smoking. But you also release toxins into the air that are inhaled by everyone around you. Conversely, when you wash your hands, you bear a small cost for the time and momentary inconvenience involved in order to gain some benefit for your own health, but your action also benefits everyone around you, at no cost to them, because they are less likely to get an infection. People have little incentive (other than as a result of altruism or social pressure) to reduce their negative externalities or increase their positive ones. It is for this reason that governments tax goods, such as cigarettes, that have negative externalities, and subsidize or encourage practices, such as handwashing, that have positive ones. Treating a resistant infection with a new, more effective drug is a good example of an action that has positive externalities. Not only is it beneficial for the patient, but there is also a society-wide benefit: the new drug may kill off a dangerous strain of bacteria before it becomes widespread.

In most of the world, either the government or insurance companies pay for new drugs, which should allow for more expensive antibiotics to be prescribed. However, payment systems are often set up such that funds for drugs must be paid out of the budget of individual hospitals, or even a single department, which puts pressure on doctors to keep costs down and not consider wider-scale benefits. This is not to say that hospitals, doctors, and patients do not care about the community as a whole; rather, each group tries to maximize the benefit to patients with a limited set of funds. Hospitals and physicians have few incentives to examine hard-to-quantify benefits that might accrue to wider society from a certain treatment.

Externalities also explain why rapid diagnostic tests are not more widely used. As we discuss in Chapter 6, such diagnostics provide one of the best ways to reduce the unnecessary use of antibiotics. In the future, they will probably allow doctors to diagnose the type of infection someone has and what antibiotics are needed, if any. Rapid diagnostic tests often cost more than antibiotics, however. For the benefit of society, it would make sense to spend this extra money to determine the infection type, since testing could reduce unnecessary antibiotic use and protect everyone against resistance. But the cost would most likely be borne either by patients or hospitals, whereas the benefit would be to the community as a whole.

A similar trade-off takes place when farmers give antibiotics to their animals to increase their growth rate, or when pharmaceutical factories release untreated waste that contains the active ingredients of antibiotics (see Chapter 7). These practices reduce costs for farmers and factories, but they encourage the development of antibiotic resistance, thereby imposing large costs on society.

Governments can address this problem in several ways, for example through taxation, subsidies, regulation, advertising, or bans on certain activities. By using taxes and subsidies to align decision makers' interests with those of society at large, many governments have successfully reduced the incidence of costly activities, such as drinking and smoking, and increased the prevalence of beneficial activities, such as vaccination. In Part II, we will look in more detail at ways to create incentives that would help solve the crisis of antimicrobial resistance.

Poor Information and Misperceptions

A person usually goes to the doctor because they have symptoms that they want to be treated. Whether it is a cough or sepsis, the underlying illness that is causing the symptoms may be hard to diagnose. Doctors must usually treat patients using empirical diagnosis, which is little more than an educated guess based on a clinical assessment of the patient's symptoms, history, and profile. This system is often not good enough to distinguish between bacterial and viral infections, or even to tell if the patient has an infection at all. If a bacterial infection is diagnosed, there is no way to determine if the bacteria are resistant or susceptible to standard treatments using empirical diagnosis, since they produce identical symptoms, and our current diagnostic tests to distinguish between them take two days to process. It is understandable that without that information, the doctor is likely to prescribe a cheaper, generic drug rather than an expensive on-patent one. This makes it very hard for more expensive drugs to show their worth, since we cannot tell when they are needed.

This problem also influences perceptions that make people unwilling to pay as much for antibiotics as for other treatments. Insurers and governments are willing to pay hundreds of thousands of dollars for a cancer treatment to extend life for just a couple of months, yet they are willing to pay only a fraction of that amount for antibiotics that would cure a patient completely. When we spoke to experts about this question, two types of misperceptions were put forward, in addition to lack of information. First, bacterial infections are quite disparate. Because symptoms from drug-resistant infections can differ so greatly, it is often unclear that two seemingly different problems have a similar origin. For example, a person who dies of ventilator-acquired pneumonia in the hospital is not regarded as having an illness similar to MRSA, TB, or a urinary tract infection, even if all of these result from infections with antibiotic-resistant bacteria. Manica Balasegaram of the Drugs for Neglected Diseases Initiative has told us that he believes this is an important reason for the lack of public support for combating antimicrobial resistance.

Second, it is often said that antibiotics would be more profitable if they worked more slowly and patients had to take them for years. However, this is not completely true; drug prices are normally determined by a drug's impact, not the length of time someone is taking it, and there are many examples of short-term drugs that earn higher profits than long-term drugs. A drug that leads to quicker healing and causes less suffering is seen as worthy of greater financial payment: it is for this reason that a single-dose cure for the virus hepatitis C is so sought after. A different aspect of antibiotics' fast action may, instead, explain their perceived lack of value. The flip side of their ability to cure quickly is that failure to use them can rapidly lead to severe illness or death. In the case of many diseases, if governments or insurance companies refuse to pay for treatment, patients and their families and friends have time to give interviews to the media to highlight their plight. Patients with bacterial infections tend to be too ill to carry out such campaigns. This lack of publicity may reduce the pressure on hospitals, insurers, and governments to pay for a cure.

Free-Rider Problem

Certain public goods, such as clean air and streetlights, can be used by everyone, whether they pay for them or not.[4] This creates what economists call a free-rider problem: someone pays for a good, and then others can use it without paying—they can free-ride. For example, everyone in London benefits from the city's extensive flood protection system, but there is no commercial way for the flood defense operators to make Londoners pay directly for this system. Residents would prefer not to pay for the system, taking its benefits as free riders. This is why governments fund public goods, such as flood defenses, highway maintenance, the police force, and the army, through compulsory taxation.

Antibiotics also suffer from a free-rider problem. Many of the products that pharmaceutical companies sell can only be used if patients are also given antibiotics. Without antibiotics, cancer treatments and surgical procedures could result in life-threatening infections, and it would be dangerous to keep patients in the hospital for extended periods. So many aspects of medicine would have to cease if we could not risk taking drugs that suppress our immune system and increase our chances of picking up an infection. However, drug companies would prefer to wait for one of their competitors to create a new, more effective antibiotic, because all companies will benefit from the new antibiotic, not just the company that develops it.

Because antibiotics provide a backbone to the entire healthcare system, antibiotic development can be thought of as a public good. But, as in the case of flood protection, private incentives are not sufficient to deliver it. For this reason, we believe there is a strong case to be made for governments to step in and provide incentives for the creation of new drugs.

The combination of externalities, cheap substitute goods, and misperceptions about the value of antibiotics have led to a society that undervalues antibiotics. When we have a market-based system for drug research, we rely on drug companies being able to make a profit in areas needing research. Societal need for antibiotics is thus out of kilter with prices. If the prices are low and the volume of sales is low, an investment is unlikely to have great financial rewards. We cannot currently rely on financial rewards as a primary motivator for new research.

Academic Disengagement

In addition to the scientific challenges of developing new antibiotics and the problem of failed economic models, a third reason for insufficient attention to the development of antibiotics is neglect by academic researchers. Universities are often the sites where early-stage research takes place—research that is crucial for innovation but rarely undertaken by industry. However, many academic researchers view antibiotics research as less exciting and new than other areas of science. And indeed, most antibiotic discovery and development programs do not differ markedly from the kind of research that Alexander Fleming and others were doing with penicillin in the 1930s.

One measure of the lower status of this field is in statistics on citation rates. Citation rates—which record the number of times an article is cited in another article—are the yardstick by which academics are measured. Citations have a huge impact on academic careers, directly influencing how academics are judged both for promotion and for new research funding. The Review on Antimicrobial Resistance found that citation rates for medical microbiology journals are lower than those in any other area of medicine it examined. Journals on infectious disease have the second lowest rates. Microbiology papers received, on average, 2.7 citations within two years of publication, while those on cancer received 3.5, and on immunology, almost 4. This relative lack of interest in the field has led to a mismatch between the number of academics working on antimicrobial resistance and the societal need for research in this area.

Short-Termism

So far in this chapter we have discussed reasons for the lack of attention to the problem of drug-resistant bacteria, and especially the factors that have hindered development of new antibiotics. But from a broader perspective, it is important to see that an infection with antibiotic-resistant bacteria is only the last step in a long chain of events. To slow the increase in drug-resistant infections, we need to focus not only on treating those who are already sick, but on reducing infection and transmission rates, reducing unnecessary prescriptions, and cutting down on the number of antimicrobials in the environment. In other words, we need to avoid short-termism.

Markov Chains

To understand the effect of intervening at different stages in the development of infectious diseases, it is helpful to make use of a model called a Markov chain that is used by statisticians and economists. Markov chains provide a way of estimating the probability of different outcomes based solely on the current situation or present state. First we will explore how the model works, and then we will apply it to the case of antibiotic resistance.

As an example, look at the chances that someone will die from a fire as a result of cooking dinner this evening. For a person to die from a cooking fire, three things need to happen. First, there has to be a fire. Then, someone has to be injured. And finally, that injury has to be fatal. For simplicity, consider just two outcomes: either someone dies from a cooking fire, or no

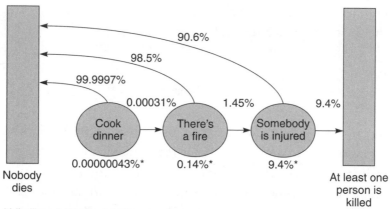

Fig. 3.1. A Markov chain for estimating the likelihood that someone dies as a result of a cooking fire. Percentages between circles indicate the probability that the event to the right occurs given that the event to the left occurs. Percentages on leftward pointing arrows above the circles show the likelihood that no fatality results from that event. Percentages below the circles show the likelihood that a fatality occurs as a result of that event.

one dies. A Markov chain lets us estimate the probability of these two outcomes at every stage of the process (without having to know previous information), as shown in Figure 3.1.

Based on statistics from the National Fire Protection Association, the probability of each stage given the occurrence of the previous one is as follows: there is a 1 in 330,000 chance that the stove or kitchen will ignite, then a 1 in 70 chance that someone is unable to avoid the fire and gets injured, and finally a 1 in 10 chance that the injured person dies. Overall, there is a 0.00000043 percent chance, or 1 in 231 million, that someone will die per meal cooked. You can expect to eat only about 24,000 dinners in your adult life, so these are pretty good odds for you, your family, and neighbors. However, once a fire starts, the odds of a death rise rapidly to 1 in 700, and once someone gets hurt by the fire, the odds of death become 1 in 10.

The beauty of a Markov chain is that you can estimate future probabilities without knowing what happened at previous stages. For example, you do not need to understand how likely it is for a fire to start to know that once it starts, there is a 1 in 700 chance that someone will die. This is why Markov chains are so useful for modeling infectious disease—we do not need to understand the exact cause of an outbreak in order to assess the options to curtail it.

A Markov chain is useful not only for estimating the probability of an outcome, but for increasing the chances of favorable outcomes while reducing the odds of unfavorable ones. By using a Markov chain to assess the effect of different interventions on a scenario's likely outcome, and combining that information with data on the difficulty or cost of these interventions, we can find the most effective way to reduce harm and increase chances of the desired result. In most cases, we discover that interventions occurring early, that is, on the left side of the chain, have a greater impact than those occurring later.

Peter Sands, former CEO of Standard Chartered Bank and now chair of a commission on global health, discussed this effect in relation to fire safety when we spoke with him in October 2016, and later in July 2017: "Deaths from residential fires have decreased massively in the last fifty years, and yet fire engines have not drastically improved, nor do we spend significantly more on fire fighting. Instead, we have improved the quality of buildings, stoves, fire retardant furnishings, and wiring to reduce fires and slow down their advance. We also have substantially increased the number of fire alarms in houses and fire extinguishers in buildings so that fires can be detected and stopped before they became a major problem, and instituted fire drills so that everyone knows the evacuation procedures. Put simply, stopping a problem before it starts is normally easier than containing it once it has commenced."

He went on to discuss the case of Grenfell Tower, a high-rise public housing apartment complex in London that was engulfed by fire in June 2017, resulting in seventy-one deaths. "The Grenfell Tower disaster illustrates the dangers of complacency. Whilst the full assessment of the causes of this disaster has yet to be conducted [at time of interview], it seems clear that inadequate and weakly enforced regulations were a major factor. Once the fire gained momentum, there was little the firefighters could do to save the people in the upper part of the building. The parallels to how we mitigate disease threats are deeply sobering."

In the case of infectious disease, the incentives, be they acclaim or a sense of accomplishment on the part of researchers, or financial rewards for companies, are often heavily skewed toward late-stage interventions, with little incentive to intervene in early-stage or preventive care. If nothing is done early on, a disease outbreak can become too difficult to contain. The 2014–2015 Ebola outbreak provides a good example of an infectious disease raging out of control. Although Ebola is caused by a virus, we have chosen to profile it because, just like a completely antibiotic-resistant bacterial infection, it has no cure; thus the Ebola crisis can provide some indications of how an outbreak of a drug-resistant infection might behave.

Ebola and the Markov Chain

The recent outbreak of Ebola started when a two-year-old boy, Emile Ouamouno from Meliandou, Guinea, came into close contact with a fruit bat. He began to show symptoms of the viral infection on December 2, 2013, and died four days later. It is now thought that the Ebola virus could have been present but undetected in West African fruit bats for some time, but it was not known then that bats were a reservoir for Ebola in this part of Africa. Better tracking might have helped us to spot this risk and prevent or curtail an outbreak.

We have a poor understanding of this type of zoonotic transmission (transmission via animals), which is the route by which many new or emerging infections pass into the human population. In part, this is because the topic falls between the disciplines of human medicine and veterinary medicine. In addition, there are not sufficient incentives to encourage the private sector to take on the job of generating this information. Because it is a public good, governments should step in to prevent this first stage of the Markov chain. But public health initiatives do not tend to be the issues that the public votes on, reducing the incentive for politicians to make them a priority.

It is much easier to stop a disease outbreak early in the Markov chain, when it has only affected a few people, than after it has spread widely. For this reason, conducting autopsies to establish cause of death, and tracking unusual illnesses, would assist in containing an emerging infection. We have known this for centuries.

Yet, even in 2013, when a child died of Ebola—an illness that was first described forty years ago—nobody recognized the illness. Nor did this happen when his mother, sister, and grandmother died of the same illness. Nor when Emile's grandmother passed it on to two health-care workers, who took the illness back to different villages before they died. It was more than three months after the first Ebola infection that hospitals first reported an unknown irregular infection to the Guinean department of health. It was another few weeks before the illness was identified as Ebola. By then, 111 people had been infected, 79 had died, and the virus had spread to four villages in three countries, in what we could describe as the next stage in the Markov chain.

There is little financial incentive for making a discovery at this stage, as there is no obvious way that spotting an emerging infection can be monetized. Industry will not invest in tracking infections. There may be not-for-profit incentives like personal satisfaction and glory for researchers who are able to prevent an emerging infection. However, since only a small number of people are sick at this stage, and it is not yet known how widespread the infection will become, the incentives are limited.

During the next six months, while the health-care systems of Guinea, Sierra Leone, and Liberia, with the help of the international community, were all working to limit the spread of Ebola, the outbreak expanded, progressing to the third stage. By September 2014, 250 new cases of Ebola were being discovered per week. Individual countries became more involved in efforts to stop the epidemic, particularly France, the United Kingdom, and the United States, who focused on Guinea, Sierra Leone, and Liberia, respectively. This aid, along with better on-the-ground support, finally led to the end of the Ebola outbreak. However, as a result of the slow global response to Ebola, the final tally of laboratory-confirmed cases reached 15,227, and 11,315 people died. The economic shock in the region was akin to 3.7 million people losing their jobs, which doubtless damaged many more lives indirectly.[5]

This crisis might not have spread nearly as much if we had had better systems of surveillance and information sharing in place, along with a strategy for early response. As this episode makes clear, our global systems and incentives often do not work to anticipate problems, but kick into gear after we are actually suffering a crisis.

The Markov Chain for Antibiotic Resistance

As discussed in Chapter 2, unless we come up with systems to curtail drug-resistant infections, it is only a matter of time before antibiotics stop working. As Jeremy Farrar, director of the Wellcome Trust, points out, "Drug resistance is very like an emerging infection. If we do not act quickly to stop it, it can spiral out of control."

Let us now return to the Markov chain model to sketch out the stages in the spread of antibiotic-resistant infections. The chain for the spread of bacterial resistance has four phases: Bacteria come into contact with an antibiotic, and selective pressures cause them to evolve resistance. People become infected by those

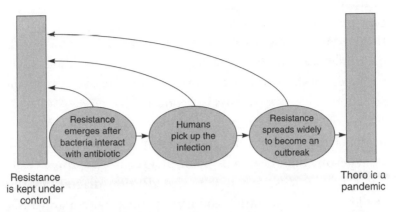

Resistance
emerges after
bacteria interact
with antibiotic

Humans
pick up the
infection

Resistance
spreads widely
to become an
outbreak

Resistance
is kept under
control

Thero io a
pandemic

Fig. 3.2. A Markov chain for estimating the likelihood of antibiotic resistance leading to a pandemic.

bacteria. Infection spreads, leading to a local outbreak. Finally, an epidemic or even a pandemic arises. Figure 3.2 shows these phases in a Markov chain diagram.

In the first stage, antibiotic resistance starts when bacteria come into contact with antibiotics. When this happens, particularly when the antibiotic dose is not high enough to kill all of the bacteria, any bacterium able to withstand the antibiotic is less likely to be destroyed; it will survive and pass the genes for resistance on to its offspring or to other bacteria via plasmids. Gradually, resistant bacteria will become more frequent. Transmission by means of plasmids can spread very quickly.

The easiest way to prevent this process is to reduce the unnecessary use of antibiotics. This can and should be done in both human and animal sectors. As discussed in detail in later chapters, we recommend use of rapid diagnostics to ensure that prescriptions are more accurate, advances in infection prevention to minimize the need for antibiotics, and reform in antibiotic manufacturing supply chains to reduce antibiotic waste.

The second stage is when resistant bacteria begin to infect people. The way to break this link in the chain is through better infection prevention and control. A study in which free soap was given to very poor households in Pakistan resulted in 50 percent lower rates of pneumonia and diarrhea. Even in high-income settings, education on proper handwashing has been shown to reduce infection rates by as much as 30 percent. Providing people with clean water and toilets is even more effective at reducing infection rates, as we discuss in detail in Chapter 5.

To prevent the third stage, a more widespread outbreak, we need better infection control. We must stop resistant bacteria from getting into the water system, and require improved cleanliness in hospitals. We should also develop diagnostics that would allow us to identify and isolate patients with highly resistant infections.

Finally, how do we keep outbreaks from turning into pandemics? At this point the political and economic incentives to find a solution are much higher than they were previously, but it is also much harder to stop these infections. Once again, infection control and diagnostics can play an important role in containing outbreaks, but new drugs are also essential for preventing this fourth and final stage in the Markov chain from occurring. To be well prepared, we should have developed useful new treatments long before we face a drug-resistant pandemic.

As George Osborne, UK chancellor of the exchequer from 2010 to 2016, told us, one of the main reasons that antimicrobial resistance is not treated as seriously as it should be is "apathy driven by the fact this isn't an immediate and obvious health crisis like the Ebola outbreak." But by the time an outbreak reached crisis levels, we would already have a huge problem that could take decades to solve. When asked what lessons we could learn from previous health crises, then director general of the World Health Organization Margaret Chan told us in November 2016 that "responses to previous global health crises have tended to be reactive. Things have to get very bad before action is taken on an appropriate scale. Think about how long it took to secure an international agreement to reduce greenhouse gas emissions."

Interventions to Prevent Antibiotic Resistance

As with Ebola and almost every other infectious disease, or indeed almost any phenomenon that occurs in a series of linked stages as modeled by the Markov chain, interventions at earlier stages are almost always easier to undertake. It is easier to stop a person from getting sick than to cure them. However, incentives tend to favor current rather than anticipated problems. Scientists have been warning people about the dangers of antimicrobial resistance for decades, but no one has heeded the warning. Now, as a result of a rise in incurable bacterial infections, outbreaks in middle-income and southern European countries, and publicity about the failure to discover new drugs over the past two decades, the public and governments are starting to pay attention. We hope that these developments and the threat of an epidemic will spur global action.

We need to consider several different types of intervention to counter the threat. Of the four main types, shown in Table 3.1 and discussed in detail in Part II of the book, the development of new drugs is the only one that cannot stop resistance from emerging. By reducing the number of antibiotics in the environment, improving infection prevention and control, and using diagnostics to help reduce unnecessary use, we can greatly reduce the number of new antibiotics needed.

But we have seen in this chapter that market forces and short-term thinking have hindered action on these interventions. When markets do not work, economists usually look to the governments to step in. In Chapters 4 through 7, we outline in detail

Table 3.1   Effects of different interventions on preventing drug resistance

	Intervention			
Effect	Discovering new treatments	Preventing infection	Reducing unnecessary use	Reducing environmental pollution
Reduces contact between bacteria and antibiotics, slowing creation of new resistant strains		X	X	X
Diminishes ability of resistant bacteria to infect people		X		
Helps prevent infections from becoming outbreaks or pandemics		X	X	
Improves response to outbreaks by treating the sick or keeping people from becoming sick	X	X	X	

the specific steps that governments should take to protect the public good by encouraging a reduction in the unnecessary use of antibiotics and generating incentives for new drug development. It is worth noting that one of the reasons governments have not intervened is that electoral cycles encourage short-term thinking. If a prime minister or president invests government resources to curtail drug resistance, they are unlikely to get huge rewards from the electorate. People generally do not vote on how well the government is dealing with a future problem, and they do not have enough knowledge of the early stages of research to make judgments. As a result, the political incentives have not been sufficient to pressure governments into action.

In 2016, then US president Barack Obama said, when talking about climate change, that while bombs and the sound of gunfire concentrate the mind, it is hard to make people focus on crises that come on gradually. "What makes climate change difficult is that it is not an instantaneous catastrophic event. It is a slow-moving issue that, on a day-to-day basis, people do not experience and do not see."

The same is true of drug-resistant infections. By the time we see the problem around us, we have already failed. Instead, we need to note the worrying trends and take action now. In Part II, we look at what changes governments can make to combat resistance, and we propose solutions that are intended to be both economically competent and politically palatable, because without taking both of these needs into account we will not succeed in our arms race against superbugs.

II

Solutions to Counter Antimicrobial Resistance

Four

Incentives for New Drug Development

Marc Mendelson is an infectious disease expert at the University of Cape Town who works at a teaching hospital. When we spoke with him in January 2017, Mendelson outlined the scale of antimicrobial resistance in South Africa, as well as the huge problems he sees daily from drug-resistant tuberculosis (TB) and HIV: "We have a massive TB prevalence in the poorer areas of Cape Town—over 2,000 per 100,000. . . . The HIV prevalence in the Western Cape mirrors the national figure of around 11 percent of the total population. So you've got 7 million people in the country living with HIV, and of course the two epidemics collide." He sees a lot of multidrug-resistant (MDR) TB, which he thinks is being transmitted directly from person to person after initially developing during the course of a long TB treatment regime. The number of cases is quite worrying: "On an infectious diseases round of ten new patients, we will see four to six with TB, and of those at least one or two will have MDR TB." An increasing number of patients have a more serious variant of the disease, known as extensively drug-resistant (XDR) TB, which has even fewer treatment options.

Mendelson notes that although new drugs are available for treating TB, they have drawbacks. If the older, first-choice drugs do not work because the patient has a resistant strain, and the newer, second-choice drugs must be used, patients often experience serious side effects: "The drugs we have to use for MDR TB and XDR TB are extremely toxic. We regularly see patients going deaf . . . and many other side effects. So there is a heavy price to pay." In his practice, Mendelson also treats patients with other drug-resistant infections, including those that do not respond to any drugs, even the carbapenems, which are the last line of defense.

Mendelson compared one of his recent patients to a well-known case from 1899 of a groundskeeper who cut his foot on a blade of grass and eventually had to have it amputated because the infection could not be controlled. Mendelson's patient had gotten a bad ankle fracture in a car accident. "He was given various courses of antibiotics and had a number of infections that were drug-resistant." The infection was untreatable, and the patient's foot had to be amputated. The infection, Mendelson noted, "resulted in the same outcome as the groundskeeper in 1899," before the era of antibiotics.

The experiences of physicians like Mendelson make it clear that we need new antibacterial drugs. But we also need to find a way to direct research and development in the right direction. In the first part of this chapter, we discuss how to change the incentive system for development of antibiotics and how these incentives should be structured. In the second part, we present a concrete proposal for encouraging new drug development. We also review some alternatives to antibiotics that are being developed to combat bacterial infections.

Role of New Drugs

Any new system for creating incentives for antibacterial research should meet five basic requirements. First, it should reflect the areas of greatest need, offering the highest rewards for antibiotics treating drug-resistant infections of greatest concern. Second, the systems of reimbursement should be efficient. Funding and fees must be reasonable, while still generous enough to spur innovation. Third, any new drugs should be affordable and accessible to patients worldwide. Fourth, incentives should be put in place to foster appropriate stewardship to reduce problems of excessive and wasteful use. Finally, any system needs to provide certainty for all parties involved that projects will not be dropped before completion. In this chapter, we compare all reimbursement models to these five standards: need, efficiency, access, stewardship, and certainty.

Need

Several times in the past few years, researchers have gained attention for their products by touting them as potential cures against superbugs. In most cases, the researchers could credibly claim that the drugs had the potential to be useful tools in the fight against drug-resistant infections. However, the part of the story they did not publicize is that because these drugs generally have not yet been in clinical trials, the chance that any of them would be approved was very low. Low success rates for all drugs, combined with few new antibiotics in development, means that the drugs we most need are in short supply.

In March 2015, Pew Charitable Trusts released a list of the antibiotics that were being tested in clinical trials worldwide or had recently been approved. There were only forty-one drugs on the list. At the same time, more than eight hundred potential cancer therapies were being tested. The Review on Antimicrobial Resistance analyzed the antibiotics on Pew's list and found that most of these antibiotics did not target the areas of greatest need. As we explained in Chapter 2, for infections that are resistant to the class of antibiotics called carbapenems, the only reliable treatment we have left is colistin, a drug that can cause kidney failure or nerve damage. Only three of the forty-one antibiotics were likely candidates for treating carbapenem-resistant bacteria (see Figure 4.1).

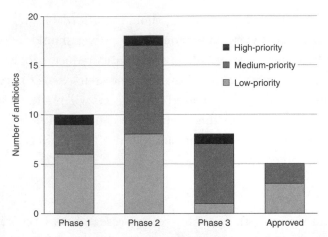

Fig. 4.1. Number of antibiotics that are in the pipeline or have recently been licensed, as of March 2015. High-priority drugs are those that have the potential to be effective against at least 90 percent of carbapenemase-producing bacteria in the UK. Medium-priority drugs target at least one drug-resistant threat labeled "urgent" by the CDC. Low-priority drugs are those not considered urgent by the CDC. (See text for more details.) Redrawn from Review on Antimicrobial Resistance (CC BY 4.0). Review's own analysis; pipeline data provided by Pew Charitable Trusts, Antibiotics Currently in Clinical Development, March 2015.

While three drugs in the pipeline might sound like a sufficient number for tackling the crisis of carbapenem resistance, the reality is that most drugs in development do not make it to market. For every antibiotic that succeeds, researchers must test thousands of molecules for antibacterial properties. Once a potentially useful molecule is identified, years are spent refining and improving it, producing it in large quantities, and ensuring that it is suitable for human consumption. Then it must be tested in animals, to assess safety and anticipate how it will perform in humans. Only a tiny fraction of molecules make it to the end of preclinical tests, in a process that normally takes more than five years. After this, clinical trials start with phase 1 studies, in which the drug is tested in a small number of healthy patients to look for unforeseen adverse effects and determine optimal dosing strategy. For antibiotics, only about a third of these drugs are deemed both safe and useful enough to keep testing. Phase 2 trials then begin: the drug is given for the first time to a few dozen to several hundred sick patients to test its efficacy. Finally, if the drug makes it through phase 2, it goes on to phase 3, where it is tested on a larger number of sick patients to gain a better understanding of the drug's efficacy and check for rare side effects. Phase 3 trials, which can take several years, recruit up to a few thousand patients. Drug companies spend about $100,000 for each patient in a phase 3 trial, and the risks of failure are high. Estimates from the consulting firm ERG on the probability of success for each phase of antibiotic development suggest that the likelihood of ultimately being approved for market at 9 percent for antibiotics in phase 1, 28 percent for those in phase 2, and 57 percent for those in phase 3.

Applying these average approval rates to the three antibiotics that were in development in 2015, each in a different phase of clinical trials, there was only about a 72 percent chance that one or more of the drugs would be approved, a 23 percent chance that at least two would be, and a mere 1.5 percent chance that all three drugs would become available. The odds of eventually discovering a drug that will tackle carbapenem resistance is high, but with only three drugs in the pipeline, the odds are not as high as they ought to be.

Moreover, even if a drug achieves regulatory approval, it may not be as effective as we hope. It is common for a drug to make it to market only to be withdrawn (due to lack of efficacy or safety concerns) or proven to be of limited benefit because it turns out to be less promising than it appeared in trials. When a drug is tested on only a few hundred patients, it can be difficult to get sufficient data on effectiveness and toxicity. We usually do not have a good sense of a drug's effectiveness until it has been on the market for a few years. Many antibiotics end up being withdrawn from the market—more than any other class of medicine. Kevin Outterson and colleagues estimated that between 1980 and 2009, 43 percent of all antibiotics approved by the US Food and Drug Administration (FDA) were later withdrawn because of problems with efficacy, safety, or low sales. While there is reason to believe that the failure rate might have been higher in the 1980s and 1990s than it is today, it is still the case that not all approved drugs ultimately succeed. So even if one of these three antibiotics makes it through clinical trials, it still might fail.

In short, the likelihood of a drug becoming available soon that is at least 90 percent effective against carbapenem-resistant infections is slim.

Of the thirty-eight remaining antibiotics in the pipeline as of March 2015, twenty, which the Review on Antimicrobial Resistance classified as "medium priority," either targeted carbapenem-resistant infections but had a success rate of less than 90 percent, or were intended primarily to treat particular infections, such as gonorrhea or *Clostridium difficile*, that are considered by the US Centers for Disease Control and Prevention to be urgent threats to human health.

The Review classified the eighteen remaining drugs as "low priority." Half of these had the potential to treat MRSA (methicillin-resistant *Staphylococcus aureus*) or to deliver an existing treatment in a new and useful way, but half promised little to no medical benefit.[1] These drugs were not novel classes, did not have a novel mechanism of action, and were unlikely to assist with current or future outbreaks of resistance.

Why were there only three potential drugs to treat the most threatening form of drug-resistant infections, which are likely to kill millions of people over the next decade, yet nine drugs for comparatively minor medical needs? The major reason why few new antibiotics are being developed is that pharmaceutical companies do not get the rewards for breakthrough research that they receive in other areas. In addition to the economic constraints to antibiotic development, and challenges of new molecule discovery, there are two particular difficulties with novel or high-priority antibiotics that limit these rewards: the patent system, and the clinical trials system.

The patent system has a different impact for antibiotics than for other drugs because their sales profile is different. Patents reward the first few years (normally about a decade) of sales for any new drug. In most areas of medicine, if a product offers genuine benefits to society, then it will make enough return on investment to cover the costs of development. For products that are not profitable, we can presume that demand, and by extension the benefit gained by society for the product, was not high. A new antibiotic, however, usually has low sales when first on the market but remains useful for decades. In most areas of medicine, the drugs used in the 1950s or 1960s have become obsolete because of medical advances. But as long as an infection is susceptible to a particular antibiotic, that drug will tend to work as well as, if not better than, the drugs we have discovered more recently.

When first on the market, a breakthrough antibiotic is likely to be a last-line drug (to be used when others fail)—incredibly important but realizing few sales. Over time, as new drugs come to market and older drugs fall victim to greater levels of resistance, the same drug is likely to move from a last-line to a first- or second-line treatment. Sales will greatly increase over time. The challenge antibiotic producers face is that patents usually expire ten to fifteen years after a drug reaches the market; therefore, the original developer loses most of the value of the drug by the time it moves to a first- or second-line treatment. The justification for this system is that an inventor should be rewarded with ten to twenty years of sales before the new product is made available to everyone.[2] If a developer has not sold a product in sufficient numbers to make a profit by that time, the reasoning goes, then its value to society was probably small. This model does not work for the sales trajectory for antibiotics, however, with low levels of early sales, followed by years of extended value.

A second factor that hinders development of antibiotics is the difficulty of conducting clinical trials for rare types of drug-resistant infections. Finding patients for these trials is more difficult than finding patients with more common types of infection. When a patient enters the hospital with an infection, treatment often has to begin right away, before doctors have been able to determine what is causing the infection and how best to treat it. In an attempt to find enough patients with these rare diseases, researchers end up enrolling patients who seem likely to have the infection of interest but who may turn out not to have it after all. In trials of drugs used to treat *Clostridium difficile* infections, only about one in four patients enrolled in a trial turns out to have that infection. This problem greatly drives up costs.

In order to truly stay on top of drug-resistant infections, it would be ideal to have new antibiotics while the current ones are still working most of the time. Both the FDA and the European Medical Association (EMA) have changed some procedures over the past decade that have made it somewhat easier to run trials for these high-priority drugs, but more needs to be done. We discuss this issue in more detail in the second part of the chapter.

Efficiency

A system for drug development should be economically efficient: generous enough to stimulate research but not wasteful. If we could spend $100 billion for every successful new antibiotic, antibiotic drug development would become a high priority—but the cost would obviously be staggering. Every dollar spent needs to be justified, since the money could also be used for other methods of reducing drug resistance, such as diagnostics, vaccines, or improved infrastructure or other areas of need. However, because antibiotics are cheap compared to the benefit they provide, it would be better to invest in antibiotic development than not, even if it cannot be done as efficiently as we might like.

In our opinion, two economic concepts should guide antibiotic development if we are to get the best value for our money. First, risks are best taken by those who have the most information about a product—in this case, the pharmaceutical industry. Second, governments are more financially stable and have greater access to cheap capital than industry. Governments should avoid borrowing money at the high rates paid by pharmaceutical companies.

Before we discuss these two concepts in more detail, we should mention a political consideration that can cause government funding of drug development to be inefficient. Electorates, for good reason, react poorly to the idea of companies receiving taxpayer money without providing anything in return. Governments are thus wary of instituting any program that could be viewed as a government handout to industry. To avoid this outcome, governments often consider less efficient systems that are harder to mischaracterize as handouts but end up costing taxpayers far more.

An example of a solution to encourage new antibiotic development that, although flawed, has the potential to succeed politically, is one in which every company developing a new antibiotic would receive a transferable patent voucher known as a market-exclusivity voucher. This voucher could be used to extend the patent on one of the company's more profitable drugs. Selling the voucher would create a payment for the company, not unlike the idea of governments giving prizes for valuable innovation; however, the ultimate cost would likely have to be higher than the prize to achieve the same amount of incentive, and such a system would drive up the cost of drugs for government, insurance companies, and anyone purchasing them out of pocket. This proposed solution has not been implemented anywhere, but since it is a system in which the transfer from government to industry is not transparent, it has gained some traction among policymakers and industry leaders.

Those with Information Should Take the Risks

Drug development is ultimately a process of managing risk. We have seen that only a small percentage of promising drugs ever make it through all phases of testing in clinical trials to reach the market. At every stage of this process, efficacy and safety data are used to assess whether the drug is good enough to continue, and difficult judgment calls have to be made about whether to abandon work on a drug.

The company running a clinical trial is the one that has the most information about the drug being tested. If a government pays for the trial (as occasionally happens with neglected diseases), however, the company does not have an incentive to cancel the program if it is not going well, since it is paid either way. This asymmetry of information can lead to a product being funded when the risk of failure no longer warrants its continuance, wasting the funder's money. We believe that, whenever possible, the organization that is dealing most directly with the prospective drug should be the one to bear the cost of failure. Government would also need to reward companies more generously if the product succeeds, since the payment would need to cover all the prospective drugs that failed en route.

Political pressure is another reason that governments are badly placed to take risk. A government may continue to fund a failing project because government ministers do not want to publicly admit to investing tens of millions of dollars in a project without results. Politically, it is far easier to continue the project, even if it is expensive, time consuming, and results in a poor product. To avoid the perception of failure, governments tend to fund too few products and to continue funding them for longer than they ought to. In the United States, a government agency called BARDA (Biomedical Advanced Research and Development Authority), within the Department of Health and Human Services (HSS), has worked to withdraw funding from drug development projects that are faltering. The agency regularly reviews the research with the help of panels from across the federal government who are politically independent, so that no one entity or individual can be blamed for a decision to withdraw funding. This is a model that could be used elsewhere.

Governments Should Invest Early

At the time of writing, the United Kingdom could borrow money over ten years at an interest rate of 1.4 percent annually, the United States at 2.5 percent, and Germany at 0.38 percent.[3] In contrast, large pharmaceutical companies must pay effective interest rates of more than 10 percent, and smaller companies must pay even more. For every dollar the United Kingdom borrows today, it must pay $0.15 in interest over ten years, while a pharmaceutical company borrowing at 12 percent would need to pay $2.10 in interest payments. The company would have to generate $3.10 over ten years for every dollar borrowed, then, just to break even. Governments would generally prefer to pay for things later, because it is cheaper than paying immediately. But the earlier a government funds a pharmaceutical company, the more incentive is generated per dollar contributed. Using the figures above, a government would generate 2.7 times more incentive for every dollar it pays the company today than for every dollar paid in ten years' time. For this reason, all other things being equal, governments and insurance groups (which can also borrow cheaply) would provide greater incentives for the development of antibiotics if they paid these incentives as early as possible. However, such payments have to be carefully calibrated against the desire that companies and not government take the risk.

Access

As we'll discuss later in the chapter, none of our proposals require that governments pick up the tab for fixing the antibiotic market, but we do believe that governments should play a role in coordinating any system so that the economic and system failures inherent in this area can be overcome. Access should be at the heart of any government policy interventions to assist the antibiotic market (financially or otherwise) in increasing the supply of new antibiotics. Prioritizing access does not mean that prices cannot be raised. What it does mean is that any intervention should include provisions that allow the poorest people in the world to have access to the new drug. High-income countries have a self-interested reason to do so: in our interconnected world, infections can quickly spread to all regions.

Looking at how other expensive drugs and treatments have been made available in low-income countries can provide a model for how to provide access to expensive antibiotics. Over the past fifteen years, various aid agencies and charities have set up programs to increase access to vaccines and to treatments for HIV and hepatitis C, and these programs have helped a huge number of people gain access to life-saving drugs, despite the high cost of these drugs in high-income countries. UNICEF for example, negotiates affordable prices for vaccines with vaccine manufacturers in low-income countries.

Publicity about how much effort drug companies are putting into accessibility could also be effective. The independent Access to Medicines Foundation was established in 2004 with the mission of helping improve access to drugs in low- and middle-income countries. Every year it produces an index that ranks the world's twenty largest pharmaceutical companies on how well they provide access to their drugs. This index was designed both to reward those companies that are good at ensuring that people have access to their products (such as GlaxoSmithKline, which topped the 2016 rankings) and to highlight those that are not as good (Astellas Pharma was at the bottom in 2016). Access to Medicines Foundation now plans to create an index called the antibiotic benchmark focused specifically on antibiotics, which will rank companies on criteria such as the amount of antibiotic research they undertake, how they place limits on unnecessary use, and whether people in low- and middle-income countries have access to these drugs.

Stewardship

One reason for the overuse of antibiotics is that pharmaceutical companies make more money by selling more drugs; their incentives are to oversell, not undersell their products. A number of groups have proposed delinkage or partial delinkage models in which research and antibiotic development would be rewarded in a way that was not linked to the volume of drug sales. A fully delinked model would remove any incentive for a drug company to drive up sales of its product.

These models raise several difficult questions. For one thing, How do you judge an antibiotic's value? In drug development, as elsewhere in the commercial world, we use willingness to pay (volume of sales multiplied by price) as a proxy for a product's societal value. If payments to companies were delinked from sales, we would need to estimate value differently. The value should reflect the need for the product as well as offering certainty to investors.

Companies currently have strong incentives to get past the regulatory hurdles, set up reliable manufacturing plants and adequate supply chains for worldwide distribution, and sell their products in as many regions of the world as possible. They produce pamphlets and literature for physicians explaining what doses to use and what outcomes to expect. A completely delinked system would either replace the pharmaceutical company with a different entity in this role, or would find another way to encourage the creation of an adequate supply chain. Those who propose a partial delinkage system suggest that there should be regulations limiting the sale of the drug, and possibly providing incentives for judicious use, but that the original innovator would still be in charge of selling the product. We believe that such a system has many benefits: it takes advantage of industry's greater risk-taking ability and expertise in manufacturing and supply, while removing the harmful incentives that push companies to oversell antibiotics and underinvest in research and development.

Incentives relating to stewardship should also do more than just stop unnecessary use. They should address quality, too. Poor-quality counterfeit antibiotics abound in many low-income countries. These products not only reduce the effectiveness of treatment, causing suffering for many individuals, but also tend to provide lower doses than prescribed treatments do. As a result, a patient can end up not taking enough antibiotics to properly clear their infection, which can spur drug resistance by allowing the least susceptible bacteria to survive and flourish.

In some areas of medicine, authorities do not see stamping out counterfeits as a priority, because they consider it better for people to have access to imperfect drugs than not to be able to afford treatment at all. That logic does not hold for antibiotics, however, where counterfeits can impose a large cost on all of society. Regulatory agencies need to crack down on the illicit manufacturing of counterfeit antibiotics as well as on their supply and sale.

Certainty

When taking a drug through the lengthy clinical trial process, a pharmaceutical company constantly evaluates scientific results to make sure the drug continues to show potential for medical use. If these results are poor, the company may decide to halt the project. But it may also do so because of low projected economic benefit and commercial value. Let's look at a situation where a company is developing two potential drugs, each with a 10 percent chance of success, for a cost of $200 million each. One of these drugs is a promising new treatment for dementia with an estimated $50 billion of discounted profit (profit after taking into account the cost of borrowing) if it is successful. It is a no-brainer to take the risk on that drug. The company would pay $200 million for a product worth $5 billion (10 percent of $50 billion). The second drug has the same cost and risk profile, but it is an antibiotic that is only expected to bring in $1 billion of discounted profit. In this case the company would be spending $200 million on a product whose worth to it now is only $100 million (10 percent of $1 billion).

This problem of the low profitability of antibiotics could be addressed by establishing government-funded incentive programs. But such programs would need to provide a high level of certainty. Joe Larsen, director of BARDA's Medical Countermeasures division, told us when we interviewed in him in November 2016 that "it is really important that if a fund were to be established, it cannot be subjected to the political whims of the day. . . . Whatever we create in terms of policy needs to stand the test of time, or industry is not going to think that it is reliable, and if it is not reliable they're not going to count on it, and if they're not going to count on it, it is really not an incentive."

When drug companies make calculations about risk, those calculations are based on the anticipated number of sales and the expected price of the drug. If we want to change the incentive system, we need to be able to convince the company that its $200 million investment will pay off. If government policy changed, and the government refused to pay the investor after it brought a successful drug to market, the company would lose money—it would already have made the investment and would have no way to recoup the money. Even a very low risk of nonpayment can greatly decrease trust and reduce the appeal of an incentive on offer.

Under the current system, investors believe the odds are extremely high that they will receive the payout they anticipate. It is worthwhile to invest time and energy in a product because it is highly unlikely that the intellectual property framework underpinning modern drug development will change. Any new incentive system must have similar guarantees and be immune from short-term change. It must not be vulnerable to being undermined by a change of government, an election, or a public whim, because if the system *can* be changed, companies will fear that it *will* change and will thus not live up to its promises.

For this reason, when asked about what kinds of incentives they prefer, pharmaceutical companies often propose highly inefficient ideas that do not involve direct payments from governments. Governments do regularly sign contracts for products that involve future funding, such as the building of a bridge or the purchase of new military equipment. But a government cannot refuse to pay after a bridge is built: if this were to happen, the construction firm would sue, and it would win the lawsuit. Pharmaceutical companies need similar guarantees for drug reimbursement. We propose that any new system guarantee that no change can be made in the terms without ten or fifteen years' notice. A combination of contract law and clear terms for payment could achieve this goal.

Solutions

Many different economic solutions have been put forward for tackling the problem of drug-resistant infections, most of which are either market-based or government-based. After discussing these two types of solutions and outlining some of the problems with them, we put forward our own proposal that includes aspects of both market- and government-based solutions. We believe that our model, which involves payment to those who come up with new antibiotics, will be able to take advantage of the benefits of both solutions.

Let the Market Decide

All proposed so-called market-based solutions involve some kind of change in the fundamental structure of the pharmaceutical market.

At the moment, in the United States, pharmaceutical companies can charge as high a price as they would like for the drugs they produce. (In most of the world, a process called health technology assessment [HTAs] is used to regulate prices.) Companies know that if they increase prices too much, there will be a reduction in demand and a loss in sales. Some people argue that the best way to solve the innovation shortage of antibiotics is simply to have higher prices for the drugs. However, higher prices would harm access and stewardship, as well as failing to provide incentives for drugs that have high value but low sales. This proposal also does not make sense, since there is nothing currently constraining companies from increasing the price of drugs other than concern that doctors will not prescribe their product or HTAs will deem them poor value for money.

Some analysts argue that companies would produce more antibiotics if hospitals could charge separately for them rather than having them included within the set fees that are charged per condition (fees that are determined by categories called diagnosis-related groups, or DRGs). At present, in many countries hospitals are reimbursed by insurance companies or governments for the average cost of a procedure or diagnosis, including any drugs that are prescribed (e.g., there is a set payment per hip replacement), with the result that if there is a problem and costs are higher than expected, the hospital must pick up the tab. This means that when someone has a drug-resistant infection, hospitals must pay the cost of the additional drugs, plausibly reducing the price they are willing to pay for those drugs. But even if removing antibiotics from DRGs did encourage hospitals to pay more for them, it would not be a good idea to do away with this system because it provides strong incentives for hospitals to keep their infection rates low.

As mentioned earlier, another proposal that some pharmaceutical companies have put forward is that of market-exclusivity vouchers. These vouchers would extend the market-exclusivity, or patent protection, of any drug; they could be used for one of the company's own products or sold to another company and could be distributed by regulators such as the FDA, under a system similar to that of priority review vouchers (vouchers that confer expedited regulatory review for certain kinds of drugs). If a company created a new antibiotic that was useful but not profitable, the company could get an additional payment through the sale of a market-exclusivity voucher. Whoever bought the voucher could use it to extend the patent of one of its drugs. The voucher, then, would essentially function as a tax on the patients and insurers needing that other drug; its price would stay high for longer because the introduction of a generic version would be delayed. We should clarify that this voucher system is not actually a market-based system, but more of a government prize; yet it would be easier for governments to implement and companies to trust because governments would not pay directly for it.

Vouchers would also vary greatly in value depending on how many drug companies had high value products near the end of their patent life, which they wanted to extend. As a result, payouts would not be linked to the quality of the proposed new antibiotic, so this approach might not encourage the development of drugs that reflect societal need. Assessed on the basis of the essential criteria we listed earlier in the chapter, this system fails in terms of efficiency and also gets low marks for need, access, and stewardship, but it does so well in the area of certainty that it should be considered if a more ambitious plan cannot be implemented.

Government or Civil Society–Based Solutions

Some groups are concerned that market-based incentive systems rely on rising prices, so they favor state interventions such as ones in which governments undertake the research directly or buy out patents because they believe it would be possible for non-profit or small firms to conduct research more cost-effectively. This is a strong argument on paper, since there are a lot of very large costs in drug development and clinical trials. Yet we are not aware of any company or individual who has taken a drug through clinical trials for significantly less money than the large pharmaceutical companies typically spend, and we doubt that it could be done.

An argument for the state to buy out the patent of any new, potentially useful drug and distribute the drug directly has more merit. This system would reflect need, because the state could prioritize the drugs that it believes are most important. It would also allow for proper stewardship, since it would remove incentives to oversell, and it would allow access, since it would be in the government's interest to make sure the drug was available to everyone. The problem with this system is one we have discussed before: pharmaceutical companies do not trust the state to pay out. Additionally, governments do not have a history of successfully managing supply chains and product distribution as efficiently as the private sector. For these reasons, we argue for an adapted version of the buy-out model as the best way to create incentives for the development of new antibiotics.

Our Proposal to Encourage Drug Development

We have formulated a proposal to encourage new drug development that meets the five requirements laid out earlier in the chapter: need, efficiency, access, stewardship, and certainty. We recommend: (1) public funding via an innovation fund that pays for and encourages early-stage research, as well as non–cutting edge research that has societal benefit but little commercial attractiveness; (2) greater collaboration among companies in conducting clinical trials, and harmonization of regulation, both of which will reduce the cost of bringing new drugs to market, and (3) so-called market entry rewards that will compensate a company for creating a product that is or will be useful. Figure 4.2 provides an outline of this model.

Public Funding for Research

Research in the early stages of drug development is best carried out by academia or governments; it is so far removed from the commercialization stage, and so risky, that companies do not want to invest in it. If the research is publicly funded, then companies can have access to it and can use the results to invest in future development. The UK Department for Business Innovation and Skills estimates that while the private return on investment from research and development can reach 25 to

Fig. 4.2. A plan to overhaul antibiotic discovery. Redrawn from Review on Antimicrobial Resistance (CC BY 4.0).

30 percent, the total societal returns are two or three times greater because of spillover effects. For example, a company that invents a new type of car engine will profit from the sales of the engine, but consumers who buy the car will also benefit, and future researchers will be able to build upon the knowledge and expertise created by the company. Universities may also benefit because companies often purchase rights to basic research from them.

We believe there needs to be more investment in very-early-stage research on antimicrobial resistance. This health problem receives less funding than others do because it is not seen as secure enough by governments or as interesting enough by scientists. Research councils and governments need to make this early-stage research a priority.

Recently the HSS's BARDA agency and the Wellcome Trust announced a new initiative to do exactly this: the Combating Antibiotic-Resistant Bacteria Biopharmaceutical Accelerator (CARB-X). CARB-X, which has its administrative base at Boston University School of Law, is a public-private partnership designed to fund the clinical trials of promising new antibiotics. The system has been designed specifically to minimize groupthink— the practice of thinking or making decisions as a group in a way that discourages creativity and individual responsibility. CARB-X's CEO, Kevin Outterson, told us in November 2016 that even the smartest, most well-meaning people can end up wasting huge amounts of money on failed research: "If we know anything from history it is that competent companies can make terrible decisions." CARB-X tries to overcome groupthink by having its projects chosen by a number of partner organizations, all with different areas of expertise and ways of thinking. It is important for scientific decision making to be undertaken by many smaller groups rather than one large one, since it is easy to make mistakes when estimating the likelihood of a product being successful. Because different groups are likely to make different mistakes, CARB-X is trying to make sure that no one way of thinking dominates their strategy; this should reduce groupthink and lead to a diversity of projects being funded. The first CARB-X projects were announced in March 2017: the partnership invested $48 million in projects being undertaken by twenty-two promising biotech companies. We look forward to seeing the results.

Additional funding is also needed to study drug combinations and dosing. Currently, most of the antibiotic dosing studies we use date back to the 1960s and 1970s. There is some evidence to suggest that we could lower resistance rates by increasing the dosage and reducing the duration of many antibiotic treatments.

Collaboration among Companies and Harmonization
   of Regulation

The cost of research during the preclinical stage, before a drug is tested on people in clinical trials, is relatively inexpensive, about $5–10 million per drug. However, this stage of research is very risky, since most projects never progress to fruition. For every drug that makes it to the market, researchers study hundreds of compounds. Reducing this failure rate, even slightly, would make drug development more efficient.

If drug companies shared information about failures and the types of research they were doing, duplication could be avoided and risk would be reduced, leading to better science and large cost savings. Most companies do not want to share this kind of information with their competitors, regarding it as not in their interest to do so. We believe, however, that greater collaboration within the pharmaceutical industry is not only possible but would be beneficial. There may be a role for government regulation in fostering such collaborative research. Some steps in this direction have already been taken. CARB-X is exploring ways to require the companies it funds to publish interim and negative results. Making these results public would help companies learn from each other's failures and avoid repeating them. We need to ensure that any forum created to share data is designed in such a way that it does not encourage more groupthink.

Further, we believe that pharmaceutical companies should work together to create clinical trial networks. The current process is very inefficient. To begin a clinical trial, a company must agree to terms with up to one hundred different hospitals. The company must then teach those hospitals about the trial's protocol, as well as when and how to enroll patients. Each step takes time, so the process of starting a trial can be slow. In addition, companies must secure laboratories to test trial results and hire researchers to oversee the trial. The majority of clinical trials run for a year or two; when completed, the network of hospitals, labs, and researchers created for the specific trial disappears. The company has no incentive to maintain such an expensive network, since it rarely has another drug to test right away. The next company with an interest in running a similar trial must go out and create, at high financial and time investments, its own network of hospitals. Letting two companies use the same network sequentially would lower expenses. A multistakeholder working group led by the Wellcome Trust, and chaired by one of this book's authors, found that the costs of a clinical trial could be reduced even further if different clinical trials shared patient control groups. The working group found that the combination of these two efficiencies (maintaining networks and sharing controls) would result in a return on financial investment more than three years earlier than is the case in the current system, with overall profitability increasing by about $175 million per drug. With this type of collaboration, many drugs that would have been unprofitable could become viable projects to pursue. The Wellcome Trust is currently working on the exact business case for a clinical trial network and how it would function. We believe that such a system would make it easier and cheaper for companies to take drugs to market in the future.

Finally, harmonization of regulation between countries would remove some obstacles to drug development. Drug regulators in the European Union and the United States have signed a number of agreements over the past decade regarding approval of antibiotics, agreeing to make standards more similar, such as the number of patients required for a trial, the statistical evaluation methods used, and even the kinds of forms to be filled out. They have also agreed to share more information confidentially. This kind of initiative should be expanded so that drug companies do not have to meet very different standards in different jurisdictions. This could be done without reducing required standards in any location by harmonizing standards that vary slightly but are equally stringent.

Market Entry Rewards

The final element of our recommended incentive package is what we call market entry rewards. This approach allows both government and industry to undertake the parts of drug development that each is best at.

In the market entry rewards system, innovators who come up with a useful new antibiotic receive a lump sum payment that is linked to the social value of the new drug. This payment is made after the new drug is taken to market and has been shown to meet a medical need that exists now or will do so in the near future. The most useful drugs receive a payment equal to the average cost of developing a new drug, in order to reimburse costs and encourage efficiency. Based on current costs, we believe that a market entry reward would need to be approximately $1.5 billion. While that figure may appear high, a billion dollars is not a lot of money in a world that spends almost a trillion dollars on pharmaceuticals every year. Of that trillion dollars, $40 billion is spent on antibiotics and $100 billion on cancer drugs, which depend on antibiotics. Yet only $4.7 billion of the money spent on antibiotics is for patented drugs. The remainder is spent on generics, which means that most of the money we spend on antibiotics does not go back to the companies that have developed the drugs.

Although the market entry reward system is an unusual way to fund drugs, it makes sense if we think about the function of antibiotics in the medical system. John Rex, one of the leading researchers in this field, argues that antibiotics provide coverage that is similar to that of a local fire brigade. We rely on antibiotics to give us peace of mind so that we can undertake tasks that would otherwise be dangerous. "When you pay the firefighters, do you pay them per fire, or do you pay them a salary? It's obviously the latter, and the importance of working together at the community level to have a fire brigade ready at all times has been known as far back as Ancient Rome."

By allowing companies to take on risk and pushing them to keep their costs down, this system is highly efficient. Although the primary goal of the payments is to spur innovation, we believe they should come with conditions: If governments pay for the creation of new antibiotics, then governments are responsible for ensuring that drugs are not overused and that everyone who needs them can afford them. Companies who commit to take a market entry reward will have to agree to manufacture the product in an environmentally friendly way and not to allow the drug to be used for livestock. Furthermore, companies must discontinue any programs that give bonuses to physicians or staff based on how many drugs they sell. Companies should be partners in the effort to stop overuse and inappropriate use of antibiotics, and they should provide the drugs very cheaply in settings where price is a barrier to good health care.

Where many in the pharmaceutical business say that the market entry reward would break down is in the realm of certainty. To generate an acceptable level of certainty that companies would receive promised funds, we believe that the criteria for payment should be simple and should be arbitrated by an independent body. The criteria should take into consideration how many pathogens the drug treats, how well the drug deals with resistance, how quickly resistance would be likely to appear, and how toxic the drug is. These criteria would determine the value of the drug, which could be converted into the amount of the payout.

Funding Considerations

All of these plans for action against antimicrobial resistance require money, particularly the market entry reward plan. Chapter 8 discusses funding and the political questions that need to be addressed in detail, but we offer some observations here.

We estimate that drug-resistant infections kill about 1.5 million people a year, as discussed in Chapter 2. In high-income countries, one percent of total health expenditure is already spent combating drug resistance, which comes out to about $20 billion a year in the United States and $1.5 billion in the United Kingdom. It makes sense for high-income countries to lead the investment to counter antimicrobial resistance. One possibility would be for countries to pay for development out of their existing health budgets; this would require about 0.05 percent of health expenditure. Alternatively, money could be raised through taxes on generic antibiotics or on antibiotics used in agriculture.

Although a market entry rewards system would provide governments with good value for their money, that does not mean that governments should fund it directly. Antibiotics are not highly profitable for the pharmaceutical industry, but they allow the industry to make large profits elsewhere: medical procedures like organ transplants and chemotherapy would become much more difficult without the support of antibiotics. We believe this justifies charging pharmaceutical companies for the cost of coming up with new drugs. A 0.16 percent tax on all drugs sold by pharmaceutical companies in the world could be used to fund the development of new antibiotics, while a slightly higher tax of 0.4 percent could fund all of the global interventions that are required to stop drug resistant infections.

Another option that was put forward by the Review on Antimicrobial Resistance was a pay or play system, whereby companies that do not invest in new antibiotics are required to pay a small investment charge to fund antibiotic development. Again, the justification here is simple: companies that do not generate the products that the entire medical system relies on, while profiting from that system's existence, should fund the required research. There are some practical questions about how to judge which companies are undertaking significant research, but these could be overcome by designating independent organizations, like the Access to Medicines Foundation, to evaluate the companies.

The Duke-Margolis Center for Health Policy, a research institute at Duke University led by former head of the FDA Mark McClellan, has been studying how a market entry reward system could work with insurance companies in the United States: each company would pay a per-patient fee that would allow patients to have access to the antibiotic as needed. The European Union's initiative on finding solutions for antimicrobial resistance has also identified market entry rewards as an important element, along with the need for greater early-stage funding of antibiotic research. International policymakers thus seem to be taking up this idea in ways we find very encouraging.

New Treatments for Bacterial Infections

In this chapter we have mostly been discussing the development of antibiotics. However, researchers are exploring a number of different treatments that could replace or reduce our reliance on antibiotics in the future. Here we briefly discuss four of these new treatments. In Chapter 5, we discuss two additional medicines, probiotics and immune stimulants, that could be used to prevent infections.

Phage Therapy

Bacteriophages, usually called phages for short, are viruses that attack and kill bacterial cells but not human cells. The technique of treating bacterial infections with phages was first discovered in the early twentieth century and was often used in Eastern Europe before the breakup of the Soviet Union, in part because antibiotics were not widely available. The Phage Therapy Center in the Republic of Georgia is still the world leader in this field. Because phages are effective on a narrower range of bacteria than most antibiotics, the therapy works best when doctors know what type of bacterium is causing an infection—information that is usually not available right away.

Little research has been done on phage therapy in other parts of the world; it came to prominence around the time penicillin and other antibiotics were revolutionizing the treatment of bacterial infections, and additional treatment methods were not considered necessary. Now that antibiotics are no longer as reliable, phage therapy is being re-examined. One challenge with this method is that phages, unlike other drugs, can evolve over time as their genetic material mutates. Drug regulations in many countries do not allow approval of a treatment that changes while it is within a patient's body; this obstacle has resulted in a lack of enthusiasm among investors.

Although there is evidence that phage therapy can be a safe and reliable way of treating bacterial infections, we are still a long way away from phage therapy replacing antibiotics on a significant scale.

Antibodies

Antibodies provide another potential alternative to antibiotics. They are proteins that the body's immune system uses to neutralize pathogens such as bacteria and viruses. Certain kinds of antibodies are widely used in cancer treatment and are being investigated for treating heart disease. But as Vu Truong, CEO of Aridis Pharmaceuticals, has stated, "The immune system evolved to fight infection, not to fight cancer or cardiovascular disease. And yet, there are not a lot of products on the market that harness immune system components to fight infections. Antibacterial antibodies are a natural first step to address that gap."

Antibodies have in fact long been used to prevent or treat certain kinds of bacterial infections, such as tetanus, botulism, and diphtheria, that are caused by toxins released by the bacteria, but they have the potential to be used for a much broader array of infections. Antibodies have several advantages over antibiotics: they may be less likely to cause bacteria to develop resistance, and they do not interfere with the desirable bacteria that inhabit the body. They could also be given less frequently: a single injection might be enough to stop a serious infection. Antibiotics, by contrast, must be administered multiple times a day.

Several antibacterial antibodies are in development at present. In two large phase 3 studies in 2015, Merck tested antibodies that bind to and deactivate toxins produced by *Clostridium difficile*. The studies showed that patients treated with the antibody had a lower rate of recurrence of *C. difficile* than those who did not receive the antibodies.

Lysins

Lysins are enzymes that kill a bacterial cell by breaking down its cell wall. They were first successfully used to treat infections in animals in 2001. At the moment, lysins have only been found to be effective against certain kinds of bacteria (gram-positive) that are not of as much concern, but with further research, their effectiveness may be extended to the more worrisome (gram-negative) bacteria. Scientists believe that bacteria will have more difficulty developing resistance to lysins than to traditional antibiotics, but until these treatments are rigorously tested, it is impossible to know if this is true.

One of the potential drawbacks of lysins is that they are large, complex proteins. Unlike most antibiotics, proteins can be recognized by our immune system, which could neutralize them and render them ineffective. This problem has not arisen in animal trials thus far.

Antimicrobial Peptides

Antimicrobial peptides are small proteins that are used by both animals and plants to defend against bacteria. These peptides have been shown to be broad-spectrum, potent, and safe antibacterial agents, and they are already being used in treatments for lung infections associated with cystic fibrosis, cancer, and several other diseases. One concern about using them more widely is that bacteria have a variety of ways of avoiding antimicrobial peptides, and if additional forms of resistance developed, the human immune system itself might lose its ability to fight against these bacteria.

Encouraging Development of New Treatments

All of these treatments show great potential for helping us win the arms race against bacteria, but at the moment they are all a long way from being successful products on the mass scale required to replace antibiotics. We encourage investors to continue to fund research into these innovative new treatments, and we believe that any new incentive structure designed to encourage new antibiotics should also apply to these alternative treatments. At the same time, new antibiotics are urgently needed.

As we have seen in this chapter, the right incentives for new drug development do not currently exist. In our opinion, it is possible to create a system that would respond to social need and promote access and stewardship, while also providing efficiency and certainty. The solutions we have discussed—increasing funding for early research and creating market entry rewards—are not unique to us; similar proposals have been put forward by US government agencies and the European Union's Innovative Medicines Initiative, and they are supported by the UK government, among others. The policy solutions to our antibiotic shortage exist; now is the time to take the necessary steps toward implementing them.

Five

Prevention Is Better than Cure

Ramanan Laxminarayan is director of the Center for Disease Dynamics, Economics and Policy in New Delhi, a lecturer at Princeton University, and an environmental economist who is an expert in the field of drug-resistant infections. When we asked him about how he became interested in resistance in January 2017, he told us about how penicillin-resistant gonorrhea spread around the world following the Vietnam War:

> Penicillin was handed out in Vietnamese brothels to prevent US soldiers from getting gonorrhea. By the end of the war, penicillin was completely useless in the treatment of gonorrhea in Saigon. That seemed unfair. It was a very focused intervention to protect soldiers, but it left behind enormous consequences for Vietnam itself, which you continue to see to this day, which is that drug resistance was jump-started by the "American war," as they called it. That penicillin-resistant strain then traveled back to the United States, and today resistance in gonococcal infections is pretty high across the board in the US. This is not just because of what happened in Vietnam [but further overuse caused the drug to become ineffective]. This seemed to me to be a problem of a global commons, which means that what happens in one country affects other countries as well, and that appealed to me as an environmental economist. It looked to me like the effectiveness of an-

tibiotics was a societal resource which anyone could access by using antibiotics, but using them depleted the stock of antibiotics that could be used by everyone else. Much in the way that you have fish in the ocean, and you can go and catch them, but once someone catches one, there are fewer fish left for others to catch. Nobody has an incentive to care about what other people are doing, or the effect on other people if there is over-fishing. What you have here is essentially overfishing in the sea of antibiotic effectiveness.

When asked about why resistance rates were so high in countries like India and China, Laxminarayan stated, "Antibiotic resistance is not just a function of use, it is also a function of disease transmission. In India, particularly, and China as well, public health and sanitation is still not sufficiently strong to prevent the transmission of infectious diseases, so that resistance, once it emerges, will spread very fast." He noted that population density is also critical for the spread of any infection. "In a typical mathematical model of disease these two parameters—population size and transmission intensity—are essentially substitutes for each other, in the sense that if you make population size greater it has the exact same effect as increasing transmission intensity." It is possible to draw a circle with a radius of 3,300 km. centered on northern Myanmar that contains about 10 percent of the world's landmass yet half its population. This region of the world has a population density far greater than anywhere else. Everything else being equal, countries in this region will always have the greatest problem with drug resistance, because once it emerges it spreads so quickly. But everything else is not equal. Other conditions exist that will exacerbate the problem: public health and sanitation rates are on average much lower, while over-the-counter sales of antibiotics remain high in some countries in this region.

Laxminarayan recommends that we begin solving the problem of antimicrobial resistance by changing how we use existing antibiotics. "For many years I thought of regulation in India, 'God that is going to be a nightmare, we're going to have to find a way around it,'" he said. "But over the years I have come to the conclusion that we are not going to find a way around it. If we are not serious about educating people, if we are not serious about getting ministers of health to get their drug controllers to make access to second- and third-line antibiotics much more difficult at pharmacies, we're throwing money at a problem that is going to keep coming back. I don't think investment in new antibiotics is really where I would put my money at all in the first place. I would put it into a massive campaign into how people think about antibiotics, in the same way that we changed how people think about tobacco. People don't smoke as much anymore, and not in public places—it is socially unacceptable to smoke in many situations now. We have got to bring that same attitude into antibiotics understanding."

The Importance of Prevention

When trying to cut back on unnecessary use of antibiotics, we face a tension between being careful not to waste these precious drugs and ensuring that everyone who needs them has access to them. The danger of over-regulation can be seen in efforts to curb the abuse of opioids: stricter regulations have led to millions of people dying in pain in low- and middle-income countries, even though a cheap and effective drug exists. Conversely, weak controls on the supply of antibiotics show the danger of under-regulation: people can easily get access to or self-prescribe these drugs, and as a result, they are becoming increasingly ineffective. When there is a conflict between these two goals we should always err on the side of access, which makes it a challenge to limit excess use. We can bypass this dilemma, however, by preventing infections in the first place—an effort that helps everyone.

Indeed, prevention is the best way to reduce the rate of infection, as we saw using the Markov chain model in Chapter 3. It is much easier to solve most problems, including that of drug-resistant bacteria, by preventing them before they start. The power of prevention was apparent during the outbreaks of Ebola and SARS, for which there was no effective treatment. These two diseases were greatly constrained and eventually defeated by preventive measures. Prevention plays two important roles in stopping drug-resistant infections. First, the more infections we can prevent, the fewer antibiotics we need to use, reducing selective pressure on bacteria and slowing the rise in resistance. Second, even if someone does develop a drug-resistant infection, effective prevention methods can prevent or greatly slow the transmission of that bacterial strain through the population.

As Val Curtis, director of the Hygiene Centre at the London School of Hygiene and Tropical Medicine pointed out when we interviewed her in December 2016, "Drug-resistant infections are a result of infection, so if people weren't getting infected they wouldn't need the drugs, and then we wouldn't get antibiotic resistance. So anything you can do to prevent the spread of infection is the primary, the most useful, and probably the most cost-effective thing you can do. Preventing infections that spread through the fecal-oral route, which is responsible for diarrheal disease and also many respiratory infections, is paramount. . . . The three best interventions to reduce fecal-oral transmission are proper sanitation, to stop people from coming into contact with fecal material; handwashing, to remove fecal material from people's hands; and clean water."

For any infectious disease to spread, a susceptible individual must come into contact with the disease-producing pathogens.

The more susceptible individuals there are, the greater the risk. And the more connections there are of a particular kind (such as coughing for some infections, sex for others) and the higher the probability that those connections will transmit the pathogen, the faster the infection will spread. The rate of spread will also depend on how many different people the infected person connects with. For example, a long period of time—more than sixty years—separated the first infection of a human with the HIV virus and the pandemic stage of HIV infection, which it entered in the 1980s. In this instance, the methods of connection were not very common, and the probability of transmission during a connection was low: a person contracts HIV by having unprotected sexual intercourse with a carrier, sharing an unsterilized needle with a carrier, being born to a mother who is a carrier, or receiving a blood transfusion from a carrier. Sexual intercourse is the only one of these connections that most people make regularly, and even then, it is much less common than something like shaking hands or sneezing and usually does not involve multiple, concurrent partners.

In contrast, most bacterial infections spread very quickly because they spread via connections that are common in many parts of the world. Many types of bacteria are spread through the fecal-oral route, where fecal material from one person ends up being consumed by another. This transfer can happen when people defecate in the open, allowing others to come into contact with someone else's feces directly, or when people drink contaminated water, or when people do not wash their hands after going to the toilet and then touch another person or a surface. People can also contract infections caused by fecal bacteria from their own bodies. The human body normally carries up to 100 trillion bacteria, and we excrete about one million bacteria in every gram of feces. Because of poor hygiene, *E. coli*, a common bacterial species in human and animal feces, can contaminate the urinary tract or bloodstream and cause a dangerous infection.

We can substantially reduce infection rates in two ways: by diminishing the number of pathogen-transmitting connections, and by using vaccination to reduce the number of susceptible people. Presently, we do not place enough effort in these areas, partly because we lack incentives to undertake early interventions, and partly because these are problems that chiefly affect the poorest and most vulnerable people in the world. Reducing infection rates is important not only to keep people healthy, however, but because doing so would greatly reduce the amount of antibiotic use. This is true no matter what the cause of the infection. Any reduction in infections, be they viral, parasitic, fungal, or bacterial, is likely to lead to some reduction in antibiotic use, since antibiotics are widely prescribed for infections of all kinds.

While we do not have good data on comparative infection rates in different countries, we do know that the proportion of people in the world who die from infectious disease is about the same as the proportion who died in the United States in the years immediately preceding the antibiotic era, as shown in Figure 5.1. These figures are in part driven by high malaria and HIV death rates in sub-Saharan Africa, yet in parts of Southeast Asia and the eastern Mediterranean, where malaria and HIV are not major killers, there are still more infectious disease deaths today than in the United States in the early 1940s. The drop in the rate of death from infectious disease in the United States and other high-income countries before the antibiotic era occurred largely through infection control and sanitation. Death rates from infection will continue to be high in countries without good preventive measures in place, even with improvements in treatment, as the example of pneumonia shows.

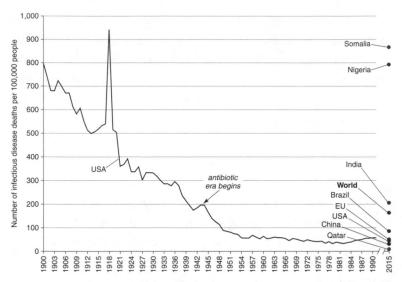

Fig. 5.1. Rate of death from infectious disease in the United States before and after the antibiotic era, compared with infectious disease death rates in the world and in selected countries in 2015. Countries shown are those with the highest and lowest rates, along with the most populous countries and the countries with the largest economies. Source: U.S. data from G. L. Armstrong, L. A., Conn, and R. W. Pinner, Trends in Infectious Disease Mortality in the United States during the 20th Century, *Journal of the American Medical Association* 281, no. 1 (1999): 61–66; 2015 data from the WHO's data repository.

Before the introduction of antibiotics, about one in six young adults who got pneumonia died. That rate rose to two in three for people over sixty-five. Presently, statistics from the World Health Organization (WHO) show that children under five have the highest risk of dying from pneumonia, but because of the availability of antibiotics, a child with pneumonia in a poor country today has a lower chance of dying than did a young adult in the early 1940s. Of the countries for which data are available, mortality rates of people with pneumonia are currently highest in Afghanistan, Nigeria, and the Democratic Republic of the Congo, with mortality rates of one in 25, 30, and 31, respectively. India, the country with the highest incidence of pneumonia in the world, has a mortality rate of just one in 105 for people who have contracted the disease. In short, because of better treatment, survival rates for people with pneumonia are higher in the poorest countries in the world than they were in the age before antibiotics. Although people who contract pneumonia have a lower chance of dying, the total number of people getting sick with pneumonia and other infectious diseases is still very high. Failure to bring down infection rates through better preventive measures means that in many of these countries, a larger proportion of people die from bacterial infections than did in the United States at the start of the antibiotic era.

The higher rate of infections in low- and middle-income countries is caused by a greater number of connections and a higher probability that a connection will result in transmittal of pathogens. When drug-resistant bacteria appear, they are more likely to spread in regions where infection rates are high. The greater number of people these bacteria pass between, and the more contact these bacteria have with antibiotics, the more likely these drug-resistant bacteria are to thrive relative to strains that are susceptible to antibiotics. It is the quick and constant transfer between people, as well as the high use of antibiotics, which has led to an estimated 200 million people in India and Pakistan carrying carbapenem-resistant bacteria in their guts. If these bacteria infect another area of the body, they are very difficult to treat using the antibiotics we currently have in our arsenal.

How to Prevent Infections

We advocate three main interventions to prevent infections: better sanitation systems, improved infection control, and preventive medicines. While the distinction between infection control and sanitation systems is not clear-cut, for the purposes of this book, sanitation includes building the infrastructure—chiefly toilets and water systems—required to reduce infection rates in the community (outside of health-care systems) and encouraging people to use them. Infection control pertains to the spread of infections within hospitals and other health-care facilities. Under the topic of preventive medicines, we discuss vaccines in particular.

Sanitation

One major obstacle to improved sanitation is that people avoid the topic of defecation. As Val Curtis says, "The problem is that nobody wants to talk about shit." Some researchers believe this disgust response evolved in humans because it protected us from the dangerous pathogens in our stool. Regardless of the reason, our reluctance to speak about this aspect of sanitation prevents us from dealing with the greatest cause of infectious disease in the world. It hinders discussion about the importance of hand-washing and proper hygiene practices, and keeps us from including the issue on policy agendas. We need to stop being squeamish, or a large percentage of the world will continue to suffer significantly more infections than are necessary.

One-third of the world's population—more than 2.2 billion people—must defecate in the open environment because they do not have access to suitable toilets. Open defecation is far more likely to result in bacteria contaminating people's food or drinking water. Further, people who practice open defecation are usually unable to wash their hands with soap afterward, increasing the probability that they will contaminate their environment, including any food they prepare, with fecal bacteria. Fecal bacteria can be easily transferred to another person with the touch of a hand, or transmitted to shared objects and counter surfaces where food is prepared. In cultures where eating is primarily conducted with hands, rather than with cutlery or chopsticks, there is an even more direct route of ingestion.

Attitudes toward defecation can also promote the spread of disease. The situation in India is a case in point. India accounts for 60 percent of the world's open defection, and it has undertaken an impressive toilet-building program aimed at eradicating open defecation by 2019. However, 41 percent of people in houses with a working toilet still defecate in the open at least some of the time, accounting for 21 percent of these households' total excrement. This reluctance to use toilets is believed to be in part because of embarrassment (people do not want others to know that they have been to the toilet) and in part because of the misguided belief that toilets are dirty, smelly indoor things. There is evidence that if people construct their own toilets, they are more likely to use them; a higher percentage of people continued to defecate in the open in households where the government directly constructed the toilet.

In addition to toilets, it is important to invest in clean water systems. Almost one in ten people globally do not have access to clean drinking water, and instead drink water containing bacteria or other potential pathogens. Humans need to drink more than a liter of water a day. The choice between going thirsty and consuming water that can make you sick is not one that any individual should be forced to make, but for many people, it is the reality. The solutions here are not theoretically complex: governments need to invest more in the infrastructure required to bring clean water to people's homes. The number of people with access to clean water has increased substantially over the past twenty years, but the process needs to be sped up. The WHO estimates that it would cost a total of $266 billion to create the infrastructure needed to give everyone in the world access to clean water. While this might sound like a lot of money, it is only 0.3 percent of the world's annual gross domestic product. It is also only a tenth of what the annual cost of antimicrobial resistance is likely to be if we do nothing about it.

Finally, education is essential, especially about handwashing. A study in the United Kingdom found that although 99 percent of people interviewed after using the bathroom in a highway rest stop said they had washed their hands with soap, electronic monitors linked to the bathroom's soap dispensers found that less than half of them had actually done so—32 percent of the men and 64 percent of the women. The results of this practice are only too clear: another study found that at any point in time, fecal bacteria are found on 26 percent of hands, 17 percent of mobile phones, 14 percent of banknotes, and 10 percent of credit cards in the United Kingdom. Data from the WHO on handwashing in nine countries reveal that rates varied significantly by country. In New Zealand, the only high-income country studied, 72 percent of people washed their hands with soap after using the toilet. Kenya had the second highest percentage, at 31 percent, while in Madagascar and Ghana the rate was just 4 percent.

Most people who do wash their hands do not do it well enough to remove all the bacteria. The US Centers for Disease Control and Prevention (CDC) recommends washing the hands for twenty seconds. Even in high-income countries, many people do not practice good personal hygiene regardless of their demographics and educational status. A study in a US university town found that while 67 percent of people washed their hands with soap after going to the bathroom, only 5 percent washed them thoroughly enough to remove all of the bacteria. The unpalatable truth is that we probably all know people who do not wash their hands after using the toilet. People of all economic levels and geographic locations need to be educated about handwashing.

Better sanitation systems would improve health and reduce the prevalence of antibiotic-resistant bacteria in the environment. We cannot emphasize enough how important sanitation is in the fight against resistance. Investments in education and infrastructure are vital to improving health care and halting the spread of infectious disease.

The Benefits of Improving Sanitation

Poor sanitation is a leading cause of childhood deaths in many parts of the world. Today, 1.7 million children under the age of five die from pneumonia or diarrheal diseases every year. Another 162 million children have a painful bacterial skin infection called impetigo, which causes blisters and swelling and results from poor sanitation practices, such as inadequate handwashing. Every eighteen and a half seconds, a child dies from pneumonia or diarrhea. To put that figure in context, at that rate, one child is likely to have died while you read this paragraph.[1] Over eight hundred will have died in the time it takes to read this book.

All aspects of improved sanitation, including working toilets and clean water, can reduce the spread of infection, but the simple act of washing the hands thoroughly with soap is one of the most effective methods. In a study on the impact of handwashing, Stephen Luby and colleagues gave poor households in Pakistan free soap and encouraged people to wash their hands after defecation and before eating, cooking, or feeding children. Compared with a control group, the incidence of illness in these households was significantly reduced: rates of disease in children were reduced by 50 percent for pneumonia, 53 percent for diarrhea, and 34 percent for impetigo. The study offered medical attention to children in both the control and treatment groups who contracted infections, so it was not possible to assess what impact these interventions would have on mortality. Yet, if the differences remained the same, as seems plausible, then, if we extrapolate to the world as a whole, handwashing with soap could potentially save about a million children's lives every year.

The benefits of handwashing are greatest in low-income areas because the infection rate is higher and people's immune systems tend to be weaker, but high-income areas also benefit from better hygiene practices. Handwashing programs have been shown to reduce respiratory infections in children's daycare facilities by 14 percent in Canada and 12 percent in Australia, and to reduce colds by 32 percent in a daycare center in the United States. A handwashing program in US school classrooms reduced respiratory infections by 21 percent, while after a US Navy training center required sailors to wash their hands at least five times a day, doctor visits for respiratory infections fell by 45 percent.

It is more difficult to quantify the benefits of installing toilets than of increasing handwashing, partly because the effects extend beyond the individual and their immediate environment. Open defecation is likely to cause problems not just for the individual doing it, but also for everyone living in the area. In addition, comparing areas that have good toilet facilities with those that do not is challenging because these areas usually differ in many other ways, as well. It is difficult to tell whether people in areas with better sanitation are healthier because of the sanitation system or because they tend to live in places that are wealthier have more effective governments.

It is also difficult to do an experiment in which one installs toilets in some people's houses and not others, partly because they often rely on infrastructure (such as a sewage or water system) that is used by the whole community. Nevertheless, some data are available. Hugh Waddington and colleagues reviewed several studies that attempted to quantify the effects of installing toilets and improving water quality and handwashing routines on childhood diarrhea. The group estimated an average decrease of 39 percent in childhood diarrhea from improvements to the sanitation system, a 31 percent decrease from the introduction of regular handwashing, and a 42 percent decrease from improved household water quality. Combining all three of these improvements would reduce the total rate of childhood diarrhea by an estimated 57 percent. Translating these percentages into mortality rates, and adding other contagious diseases, we can easily see that improvements in sanitation would result in a profound reduction in child mortality along with reducing illness in adults and preventing resistance.

Infrastructure Needs
Improving sanitation infrastructure is an important step toward combating antimicrobial resistance. The international community signaled its willingness to commit to this goal in 2015, as one of the 2030 UN Sustainable Development Goals (SDGs) that were adopted by 164 countries. Goal number six specifically deals with sanitation and hygiene. The document lays out the following objectives for this goal:

6.1 By 2030, achieve universal and equitable access to safe and affordable drinking water for all.

6.2 By 2030, achieve access to adequate and equitable sanitation and hygiene for all and end open defecation, paying special attention to the needs of women and girls and those in vulnerable situations.

6.a By 2030, expand international cooperation and capacity-building support to developing countries in water and sanitation-related activities and programmes, including water harvesting, desalination, water efficiency, wastewater treatment, recycling and reuse technologies.[2]

6.b Support and strengthen the participation of local communities in improving water and sanitation management.

Countries committing to these objectives must take into account the population growth that is expected to occur by 2030, growth which will be greatest in the parts of the world that are currently lacking proper sanitation facilities.

The World Bank estimates that the cost of extending basic sanitation and water services to resource-poor countries will be about $28 billion per year between 2015 and 2030, or about 0.1 percent of the combined gross domestic product (GDP) of the 140 countries included in its analysis.[3] Though some parts of the world will have to spend a higher percentage of GDP to achieve these goals (countries in sub-Saharan Africa will have to spend about 0.6 percent of GDP), most high-income countries will not need to spend any money at all. The cost of maintaining services after they are built will be significantly more, at about $114 billion per year globally—approximately 0.1 percent of the total world's GDP, or $15 for each person, if equally divided among all of the world's inhabitants.

Policies to Promote Behavioral Changes

In this section we focus on how to institute policies to change behavior in a way that can prevent infections from occurring. In Chapter 6, we discuss a different aspect of behavioral change: how to encourage people not to take antibiotics when they do not need them. Both are important for reducing the spread of antimicrobial resistance.

Behavioral interventions can be very cost-effective; public campaigns can change behavior in a meaningful way that can substantially improve health, for very little money. Many such campaigns do not have a sustained impact on behavior, however; these failed interventions are in turn a very poor use of precious resources that could be spent in other ways to improve health. In addition, it can be very difficult to determine which interventions have been successful, making it hard for policymakers to point to successful campaigns to justify this type of intervention. Normally when economists seek to measure the impact of a certain policy, they try to establish a counterfactual—what the world would look like if the policy had not been instituted—and compare that outcome with the outcome that resulted from the policy. This is a difficult task to undertake for behavioral interventions, since other factors usually change in addition to those caused by the policy. For example, a school might teach students about the importance of handwashing, give them free soap, and ask them to discuss handwashing with their families. A year later, if the researchers want to test the benefits of this specific campaign, they have to find a way to measure how handwashing rates changed after the intervention. But other things may have changed during this time: attitudes toward handwashing might have altered in society as a whole, or the community may have become wealthier and spent more money on soap, or another local initiative might have been going on at the same time. It is quite possible that handwashing rates increased substantially because of some factor other than the education program. This problem can be addressed by comparing before and after rates in a similar community, where all factors except the policy in question were also in play. This cannot be done, however, in the case of very large public-health campaigns, such as those that target an entire country, since no other country is similar enough to use for comparison. This difficulty in measuring the impact of behavioral change results in underfunding. Policymakers often focus on what is quantifiable, and those who fund interventions may balk at introducing a behavioral-based intervention whose effect is uncertain. Such interventions can also be harder to defend politically because the results are not obvious.

Despite these problems with quantification, there is a wealth of literature on what characteristics are most important for a behavioral intervention to be successful. For one thing, encouraging one-off behaviors is more effective than changing ingrained habits. It is much easier to convince someone to get a test for HIV, for example, than it is to convince them to always use a condom during sex, or to wash their hands. Any campaign should thus focus on the minimal behavioral change necessary to achieve the desired effect. Second, campaigns need to be continuous. Successful advertising campaigns start with a message that appears at least eight to ten times a day per outlet, to convince people of its importance. After the message starts to stick (which can take several years), an ad's frequency can be scaled back but should never cease entirely, since people are likely to forget the message.

Another important point is that the best campaigns are based on emotions and peer pressure rather than facts. Most people find statistics and facts dreadfully boring. There is often a disconnect between policymakers and the people they are trying to influence, where those designing behavioral campaigns think, "If only others knew the bad effects of their behavior, they would stop doing it." People find it much easier to ignore facts such as "smoking kills" than to ignore social stigma. Many campaigns succeed by creating a sense of disgust. One such campaign, which aimed to portray people who did not wash their hands as disgusting, was run in fourteen villages in India in 2011–2012. This campaign focused on a character named Super Amma ("super mom"), a good mother and likable character who washed her hands carefully and always made sure her children were clean. Super Amma was contrasted with a dirty, dislikeable male character. People in the villages watched skits and an animated film and were asked to make a pledge to wash their hands. The campaign was very successful; six months after it ended, the rate of handwashing with soap was 31 percent higher than in the control group, which had not received the intervention. Because social norms play such an important role in establishing behavior, such campaigns work best when they involve community engagement. If people think the whole community is undertaking a certain type of action, such as handwashing, then they are more likely to want to do it themselves. It is also helpful to use striking or crude imagery or rhetoric to make the message memorable and capture attention.

The best campaigns recognize the obstacles to behavioral change and work to overcome them. For example, people are often embarrassed for others to know that they have defecated. In communities where people traditionally defecate in the open rather than using a toilet, they may refuse to use public toilets, since their dignity or privacy is violated if their neighbors know that they are going to the toilet. In such situations, toilet campaigns seek to overcome the social stigma surrounding their use. Campaigns can also capitalize on periods of transition, which are often the best times to break habits. Teaching a new mother about the importance of washing and offering her soap to do so can be an incredibly successful way to improve family hygiene.

One of the problems with hygiene campaigns in low- and middle-income countries is that different nongovernmental organizations (NGOs) may run campaigns to reach the same or similar goals, yet none is prominent or long-lasting enough to achieve the desired change, and the varied messages crowd each other out. For this reason, we believe there is a need for a coordinating body that can make sure different campaigns are working together and are presented in a medium that people can access. Campaigns encouraging handwashing, for instance, should be aimed at the poorest people, who in many low- and middle-income countries have low rates of literacy and do not have access to televisions or the internet. Furthermore, campaigns should not promote behaviors requiring services that are not available. A campaign promoting toilet use in areas where such facilities are not easily accessed is not just wasteful of resources; it is cruel to convince people that they are damaging their own health or that of their children when it is nearly impossible for them to change their behavior.

We recommend that this coordinating body be created as a specific behavioral change unit (either as a new body or as part of an existing one). This proposed body would not need to run campaigns directly but would coordinate organizations already funding campaigns. The group should work to improve understanding about what types of campaigns work best for preventing resistant infections, and in what environments. Such a body would create an evidence base and would link projects so that funders do not duplicate work, but instead coordinate and work to fill the gaps in campaigning. Creating such an entity would be relatively inexpensive—at most a couple of million dollars a year—and would be one of the best ways to stem the spread of drug resistance.

Infection Control

It is much easier for bacteria to get from one host to another when the hosts are in close proximity. This is why urbanization greatly increased the number of people killed by infections, and why outbreaks of infectious diseases are such a problem on farms, where animals live closely together.

Another environment where people are in close proximity is the hospital. In addition to housing large numbers of people, hospitals have other risk factors for the spread of disease. Patients are likely to have weakened immune systems or open wounds which make it easier for them to contract an infection, or they may come into the hospital with an infection, possibly a drug-resistant one. Doctors, nurses, and aides move from patient to patient. They assess and dress wounds, clean fecal material from bodies, clear mucus, and look in patients' mouths and other orifices. All of these behaviors make it much more likely that the health worker or the equipment they are using will pick up and transmit potentially dangerous pathogens. The environment is ideal for the development of antibiotic-resistant bacteria since many people in hospitals are already on antibiotics (about one in three in the United Kingdom). People on antibiotics are at high risk for the bacteria in their guts to become resistant; any infection picked up from their fecal material can then spread the drug-resistant bacteria to others.

Infection control measures are intended to prevent the spread of infections within hospitals. We recommend four primary interventions to improve infection control: Medical professionals should wash their hands regularly, hospital rooms and equipment need to be properly cleaned, diagnostic devices should be used to monitor individual patients' risk, and surveillance should be carried out so that patients with drug-resistant infections can be isolated.

As we have already discussed, improper handwashing is a major contributor to the spread of infections. Here we look at the crucial role handwashing can have in preventing hospital infections. As many as 40 percent of the drug-resistant infections in US hospitals spread because health workers do not wash their hands as often as they should. A 1992 study estimated that a 28 percent increase in handwashing rates led to a 22 percent reduction in hospital-acquired infections. These figures suggest that increased handwashing is the best intervention there is for reducing transmission rates of pathogens. Making this change is not as straightforward as it might seem, however. In low- and middle-income countries, approximately 35 percent of hospitals do not have soap, and a similar number lack clean running water. Even when these features exist, they are often poor in quality or hard to access. Studies show that increasing the numbers of sinks on a ward can increase handwashing rates among medical personnel.

Even in high-income countries with excellent facilities, however, handwashing rates in hospitals are lower than they should be. This is not because health professionals do not know the importance of handwashing but because, just like the 95 percent of people who do not wash their hands as well as they ought to, they are not in the habit. To fully comply with handwashing directives, health professionals often have to wash their hands more than one hundred times a day. At fifteen seconds for each handwash, and ignoring the additional time it also takes to walk to the sink or dry the hands, that adds up to twenty-five minutes a day. Considering the huge time pressures placed upon most health professionals, cutting corners is understandable. In addition, we often believe we have washed our hands more often than we really have. Handwashing is controlled by habit—we do not make specific decisions about it or remember having done it. This, along with social pressure to conform to rules, results in people claiming that they in fact do wash their hands adequately. Inability to recognize one's own contribution to the problem, of course, makes it harder to change.

As discussed earlier, this problem is compounded by the fact that changing behavior is very difficult. A number of good campaigns have managed to increase compliance rates to between 60 and 80 percent, yet these rates diminish over time. We need to constantly remind health professionals to wash, and build facilities that make it easier for them to do so. Hospitals must undertake frequent campaigns to sustain high levels of handwashing compliance.

Cleanliness is the second important aspect of the hospital environment that affects the spread of infection. If hospitals are not cleaned properly, bacteria can be transferred by guests, food carts, doctors' hands, or any other such mechanism. Devices, from stethoscopes to an IV line, can carry bacteria from patient to patient. Hospitals need powerful cleaning methods to keep pathogens from spreading through these means. In one study John Boyce and colleagues tracked cleanliness in twenty hospital rooms by testing the same spot for bacteria both before and after cleaning. In each room they tested the bedside rails, TV remote control, bed table, toilet seat, and bathroom grab bars, and found that 24 percent of the surfaces were still contaminated with MRSA (methicillin-resistant *Staphylococcus aureus*) after cleaning had taken place. They also noted that TV remotes and bed tables were the worst offenders for retaining bacteria. Providing housekeepers with feedback resulted in a greater number of surfaces being cleaned well.

In addition to handwashing and cleaning, we can improve infection control by assessing a person's risk of contracting an infection. About 30 percent of people carry staphylococcus bacteria on their skin, typically in their nose, mouth, genital, or anal areas, where it is completely harmless. Because these bacteria can cause an infection if they enter the body through a break in the skin, surgeons often prescribe an antibiotic shortly before any operation takes place. If the person has an antibiotic-resistant strain of staphylococcus (the most common of which is MRSA), this treatment may not be effective. For this reason, hospitals now often test patients before surgery and postpone the operation until any antibiotic-resistant bacteria can be cleared. One study found that undertaking such precautionary steps reduced the rate of surgical-site MRSA infections from 0.23 percent to 0.09 percent. We suggest that this approach be extended to other types of infections.

Finally, improved diagnostics and surveillance could make it easier to isolate patients who carry resistant infections. If we know that someone has a drug-resistant infection, placing them in an isolation unit can be the best way to keep the infection from spreading. Diagnostics could aid in maximizing the use of hospital resources by helping to ensure that no patients are isolated unnecessarily and that all patients who ought to be isolated are in fact isolated.

Vaccines and Other Preventive Medicines

In addition to the preventive measures just outlined, there are several kinds of preventive medicines that can be used to keep people from contracting infections. Vaccines are the most important of these; two others that show promise are probiotics and immune stimulants.

Vaccines

Vaccines work by stimulating the immune system to generate antibodies against a particular kind of pathogen (usually a bacterium or virus, but sometimes a fungus or parasite) so that the body will be ready to fight that pathogen off later if it needs to. Four types of vaccines can play an important role in preventing drug-resistant infections: (1) vaccines that target the most common types of antimicrobial-resistant bacteria; (2) vaccines that target bacteria responsible for hospital-acquired infections; (3) vaccines against viral infections; and (4) vaccines used for farm animals and fish.

Vaccines against a large number of bacterial infections are available, including those for whooping cough (pertussis) and pneumococcus, a bacterium that can cause pneumonia and meningitis. Where such vaccines exist, we need to make sure they are widely used. We also need more research to create vaccines in areas where they do not currently exist. A recent study estimated that a universal pneumococcal conjugate vaccine (one that could treat all strains of pneumococcus, rather than just three or four) could potentially prevent 11.4 million days of antibiotic use per year globally, in children younger than five. These children would otherwise have been treated for pneumonia (from *Streptococcus pneumoniae*). Vaccines against *Clostridium difficile*, carbapenem-resistant bacteria, the gonorrhea bacterium, and certain strains of *E. coli* could all greatly reduce our reliance on antibiotics by preventing people from developing infections in the first place. We discuss below some of the obstacles to vaccine use and development.

Vaccines could also be more widely used for hospital-acquired infections. Some vaccines only provide immunity for a few months or even less, because the bacteria they target can evolve to evade the vaccine. It does not make sense to give these vaccines out to the general population because of the need for frequent boosters to maintain immunity. However, patients who have to go into the hospital are at much greater risk of picking up a resistant infection, and as their immune system may already be compromised, they may have more difficulty fighting that infection. To reduce the number of hospital-acquired infections, more work should be undertaken on the health benefits of vaccines for people entering hospitals for elective treatments.

Vaccine resistance, like drug resistance, is a concern, but it is much less of a problem. Bacteria and viruses can evolve to circumvent people's immune systems. This started happening long before humans invented vaccines; the reason that influenza and rhinovirus (the common cold) are so frequent is that new strains of the viruses keep evolving. Because vaccines increase the number of people who are immune to a pathogen, they can speed up this evolution in a process known as "vaccine escape." This phenomenon is much rarer than antibiotic resistance, however, because vaccines, unlike antibiotics, do not put selective pressure on microbes in a person's body (such as bacteria in your gut). Instead, a pathogen would have to evolve spontaneously in one person and gain an advantage by the fact that it could be passed on to a higher proportion of the population. An additional difference between antibiotic resistance and vaccine escape is that in the case of vaccines, the only selective pressure comes from a pathogen's inability to infect a person, meaning that overuse of vaccines is not a concern. Vaccine escape is most likely in pathogens that evolve quickly, like influenza, or those with many different strains in existence, like *Streptococcus pneumoniae*, which causes some forms of pneumonia.

The third kind of vaccine needed to reduce the spread of drug-resistant bacteria is vaccinations against viruses. Why should this be? The reason is that antibiotics are often prescribed for infections that are actually viral, not bacterial. For example, one common form of diarrhea, especially common in children, is caused by a rotavirus, but since the symptoms are nearly identical to those caused by some bacterial infections and we lack the rapid diagnostics to distinguish between the two, people are often given antibiotics to stop this infection. If the rotavirus vaccine was more widely used, we could reduce the overuse of antibiotics.

Finally, as we discuss in Chapter 7, about half of all antibiotics are used for farm animals and pets, rather than for people. As we will see, Norway was able to markedly reduce its use of antibiotics in farmed fish by introducing a vaccine program. More research is needed to create vaccines that will reduce our reliance on antibiotics in agriculture and aquaculture. Because it is easier to get regulatory approval for vaccines in animals than in people, this is an area where progress could happen rapidly.

At present, not enough research is being done on vaccines, and for those vaccines that are available, usage is lower than it should be. One challenge for pharmaceutical companies is that vaccines have to be made in very large batches, requiring producers to estimate need and to take the risk of producing too much or investing in an area where there is insufficient demand. Because vaccines have large positive externalities—their benefits accrue to the community as a whole, not just to the person receiving them—it is appropriate for production to be subsidized. We believe these subsidies could be best provided through an advanced market commitment. Government or NGOs would commit to buying or subsidizing certain amounts of a particular vaccine, after consulting with health professionals. Vaccine manufacturers would then be able to sell vaccines in enough quantity to recoup their investment and manufacturing costs. This process would also allow health professionals to decide which vaccine is best for a given illness, driving innovation in a collaborative environment. For example, vaccines which prevent the same illness often have different advantages and disadvantages: one vaccine might need to be kept cold, which can be difficult to do in countries with unreliable refrigeration, while a second vaccine might require two or three doses, which poses its own challenges. If a central board provided subsidies in proportion to the sales of each type of antibiotic, then the system would give the greatest reward to companies with the most requested product. This type of system was introduced in western Africa by an organization called Gavi, the Vaccine Alliance, and it has resulted in a large increase in rates of pneumococcal vaccination.

Even when vaccines exist, supply chain difficulties, limited resources, and low prioritization in some health-care systems often make getting access to them difficult. To further complicate matters, people often resist using them. Such "vaccine hesitancy," as it is called, is a well-documented problem; people may refuse vaccines because they are currently healthy and distrust the medical advice advocating them or because they cannot afford them. Good vaccines are available for preventing pneumococcus and rotavirus infections, but a 2017 WHO report found that these were only used in 42 percent and 25 percent of the global population, respectively. This is particularly problematic in low-income countries, since vaccines are expensive, and people regard vaccines as less essential than other forms of medicine (because vaccines are administered when people are healthy, so there is less concern about access). On the other hand, vaccination rates tend to be high in high-income countries, where there is a long history of vaccine use, and strong institutions that encourage it. The United Kingdom made the vaccine for smallpox mandatory for all children in 1853, and it now has one of the most extensive publicly provided vaccination programs in the world. This early and successful use of vaccines in Britain built social and institutional support for immunization. Several public-private initiatives, such as Gavi, PATH (formerly Program for Appropriate Technology in Health), the Bill and Melinda Gates Foundation, and the Clinton Health Access Initiative, use varied funding strategies to bolster development and access to vaccines in low-income countries. These strategies often involve funding the research costs of new vaccines, while the recipient pays the marginal cost of supplying the individual immunizations.

Probiotics and Immune Stimulants
Probiotics are another type of preventive medicine that has the potential to reduce the incidence of drug-resistant bacterial infections. Probiotics are medicines containing live bacteria, and sometimes yeasts, that can help restore the natural balance of bacteria in a person's gastrointestinal system. Often, when someone takes antibiotics, the beneficial bacteria in their gut are killed, but some drug-resistant bacteria may survive. These bacteria then have a huge advantage; they can colonize the gut before the "good" bacteria are able to re-enter the system and fill the vacancy. If people take probiotics after they finish their course of antibiotics, this may help the gut flora replenish itself with the good bacteria instead of the drug-resistant ones.

Results from the first large trial of probiotics, published in 2017, provide evidence that this medication could greatly reduce the risk in newborns of sepsis, a severe and often fatal infection. In the study, by Pinaki Panigrahi and colleagues, probiotics were given to 2,278 infants in India for seven days. Another identically sized group received a placebo. Both groups were then monitored for sixty days. Of the children given a placebo, 27 were admitted to the hospital with a confirmed case of sepsis, compared with just 6 of the children in the treatment group. More research is needed to confirm these promising results, but in time, probiotics may prove to be an effective way of reducing infections in newborns, as well as in other people at risk of infection.

Immune stimulants are medicines that can stimulate a patient's immune system to help stave off bacterial infections. They include treatment with antimicrobial peptides or antibodies, as mentioned in Chapter 4, and they can be given to patients who are believed to be of great risk of getting an infection, such as those going into the hospital. It is not yet clear what the best way to stimulate the immune system is, and whether these medicines should be used as treatments or as preventive measures.

Speaking in 2014 about the Ebola crisis, then president Barak Obama stated, "The best way to stop this disease, the best way to keep Americans safe, is to stop it at its source in West Africa." This message is relevant for drug-resistant infections as well. Taking appropriate steps to stop drug resistance from emerging will be much easier than trying to deal with the crisis after it erupts. That is why we need to improve sanitation systems, tighten infection control in hospitals, and make better use of preventive medicines.

In other chapters, we discuss at length the economic and political failures that have resulted in an unacceptable level of drug-resistant infections, and offer workable solutions to overcome these problems. When dealing with the two most important areas of prevention, which are sanitation and infection control, the solutions are straightforward; the difficult part is changing people's behavior and quantifying improvement. The private sector will not build the sanitation systems the world needs or invest in education to change hygiene behavior, and public pressure is not sufficient. Government financing and regulation are necessary to meet these challenges. Governments must determine what evidence-based decisions are best for their people and work to put them in place.

Six

Reducing Unnecessary Use of
Antibiotics in Humans

John Rex, a physician and a leading international expert on antimicrobial resistance, spoke with us in May 2017 about how rapid diagnostics could be used to curtail the unnecessary use of antibiotics. He began by invoking the television series *Star Trek*, where doctors on board the spaceship used a handheld device that could instantly scan and diagnose medical conditions. "It's so easy to [imagine] the *Star Trek* scanner vision of a diagnostic in which Dr. McCoy or Dr. Crusher rips out their 'tricorder' and waves it over someone and says 'You have Arcturian fever,'" he said. Unfortunately, such a device exists only in the world of science fiction.

Rex noted the challenges most doctors face: "Right now it's easier for me to give you a prescription for an antibiotic for your sniffles than it is to go to the trouble of performing a diagnostic [test]. . . . Any diagnostic takes time (unless it's Dr. Crusher's scanner), and time is precious in the clinical setting. Saying: 'take this' and 'bye'—you're done. I don't have to do anything else with you. Whereas, creating a desire to use a diagnostic requires everyone to participate." Another of the problems, explained Rex, is that patients expect to receive medication when visiting a doctor, but a diagnostic test might show that a prescription for an antibiotic is not the appropriate treatment. Rex continues:

> The patient has to be willing to say: "All right, I've gone to all this trouble to go to a doctor for relief and

the doctor's telling me that there isn't anything that's going to help—I took an afternoon off to bring you my three-year-old who's not been sleeping at night and you're telling me you can't do anything for me? I want a prescription." Also, the prescription somewhat validates the fact that the child was sick enough for the mum to take the afternoon off. "I had to go to the doctor and they gave me a prescription for an antibiotic for his ear infection because it was so bad." Everyone understands that. They don't understand: "I went in to the doctor and they said there wasn't anything they could do, so why did I go in to begin with." That's a real social problem. The employer has to deal with that. The mum really desperately wants relief, and you've not given her anything. So you have to educate the mum, you have to educate the doctor, you have to educate the entire population that it won't help. It's actually flipping the current narrative to say that antibiotics can actually hurt you if you don't really need them.

Why don't we have a device like Dr. Crusher's scanner yet? "Diagnostics have the potential to revolutionize medicine and the use of antibiotics," said Rex, "but they face commercial and scientific challenges at least as great as the antibiotics themselves."

Despite the difficulties, there are a number of exciting developments. In December 2016 we spoke with Jonathan O'Halloran, chief scientific officer of QuantuMDx, one of several companies that is developing a rapid diagnostics device: "It's not that we're creating a new diagnostic test to fit within what we have right now. Another box to put in the lab—that's not what we're talking about at all. We're looking at a change in the whole market. A complete paradigm change, which is difficult."

When asked how long it will be before it is routine for a person (at least in a high-income country) to be tested before receiving an antibiotic, O'Halloran answered: "Within five years that will happen, and I think it will happen in the pharmacy. . . . It won't just [determine if the infection is] viral or bacterial. It will start there, but it will go on to [determining] whether it's a fever. Then what kind of fever is it—is it a cold or is it flu. What kind of flu. . . . With a pharmacy model, where you've got access to your doctor, you're in the pharmacy, you can get the diagnostic and then the drug in one visit. . . . I think it's a great way of reducing primary health-care costs." It might not quite be Dr. Crusher's scanner, but the way we diagnose infections looks likely to change. O'Halloran thinks that if it does not, we are facing a lot of trouble. Referring to the development of more and more potent drugs to treat antibiotic-resistant infections, he says, "Right now we are building bigger bombs to kill our enemy, but we still don't know who or where the enemy is to use them. If we knew where they were, we could choose the right bomb. The right bomb will ensure none escape and we will contain them. Only when we contain them can we start to eradicate them and win this war. Knowledge is power. Mindless bombing exacerbates the problem." It is an arms race we cannot afford to lose, and greater precision would help.

In addition to peering into the future, it is also useful to consider the history of antibiotic use and diagnosis to understand why antibiotics are overused. Progress in many areas of medicine has been astonishing during the past century. Previously impossible complex surgeries can now be undertaken, many cancers can be treated, and heart disease often prevented. This progress has included radical improvements in diagnosis as well as treatment. In the case of cancer, advanced magnetic resonance imaging (MRI) and computed tomography (CT) scanning enable clinicians to see with unprecedented clarity inside the human body, in three dimensions. The list of such examples goes on and on. In some areas of medicine, however, progress seems shockingly slow. Diagnosis of infections is one.

Even in wealthy countries, most doctors use the same diagnostic approach today that was standard in the 1950s. In this approach, called empirical prescription, the doctor assesses the patient's physical symptoms and then makes an educated guess about the appropriate treatment. The doctor must decide, first, whether the patient has an infection; second, whether it is bacterial or viral; and third, if antibiotics might be effective. Inevitably, this process leads to many unnecessary or inappropriate prescriptions. A 2016 study by the Pew Charitable Trusts found that of 154 million annual outpatient visits at which a prescription for antibiotics was given, around 47 million (just under one-third) of these prescriptions were unnecessary. Other organizations have come up with similar estimates, suggesting that 20 to 50 percent of antibiotic use in humans is inappropriate.

Although empirical prescription is a flawed process, in many parts of the world the situation is worse, and leads to much higher rates of antibiotic use. Antibiotics are still sold in shops by untrained, nonmedical staff to anyone who wants to buy them. This situation is not easily remedied. In some countries, particularly in urban areas with more trained clinicians, governments and regulatory authorities should limit over-the-counter sales. However, severely regulating such sales could do more harm than good in rural areas of low-income countries, since it would keep antibiotics from people who really need them. We have to ensure that those who need antibiotics can get them, while at the same time limiting excess use.

In the following sections we review the costs of overusing antibiotics and then explore several ways in which we can reduce unnecessary treatment: raising awareness among both physicians and patients, improving diagnosis, changing dosing recommendations, and increasing surveillance and data collection. Finally, we return to the benefits of rapid diagnostic tests and the obstacles that have hindered their development.

The Costs of Unnecessary Antibiotic Use

Overuse or misuse of antibiotics leads to both economic costs and health costs. If an antibiotic is prescribed when it is not needed, someone, often the patient, pays an unnecessary expense. Even if an antibiotic is needed, an ineffective one might be prescribed, delaying the right treatment and causing further pain and suffering, and in some rare cases a risk of death.

Another possible cost to the patient is that each course of antibiotics, whether effective or not, increases the chances of drug-resistant bacteria proliferating in the body, especially in the gastrointestinal system. (Contrary to popular belief, the patient does not become resistant; the bacteria do.) Bacteria in a patient's body are likely to be resistant for a short period after a course of antibiotics. Often these bacteria dissipate over several months without causing a problem, but occasionally they cause a drug-resistant infection to develop. Paul Cosford, medical director of Public Health England, told a committee of UK members of parliament in November 2016, "We've got good evidence that if you or I have a course of antibiotics now, within three months our risk is three times [greater] to get a resistant infection of some sort because we've had the antibiotics affecting all the organisms in our bodies. If you're a child you're twelve times more likely to get a resistant infection in the three months after a course of antibiotics."

There is also growing evidence that taking antibiotics can affect the overall population of bacteria in the body, known as the body's microbiome. Mark Woolhouse, professor of infectious disease at the University of Edinburgh, explained to the *Guardian*: "[Antibiotics] can change the ecology of the gut, a bit like using a pesticide in a rich woodland." Killing off some species provides an opening for others to thrive. Martin Blaser, director of the Human Microbiome Program at New York University, raised awareness of the problem in his 2014 book *Missing Microbes: How the Overuse of Antibiotics Is Fueling Our Modern Plagues*. Blaser examined in detail how the overuse of antibiotics (along with other practices, such as cesarean sections) can change our microbiome, increasing the likelihood of developing diseases such as type 1 diabetes, obesity, and asthma.

Unnecessary antibiotic use also leads to wider societal costs—costs that individuals rarely recognize when they take an antibiotic. These costs come from the negative externalities of antibiotic use, as discussed in Chapter 3: when any antibiotic is used, it increases the chance that drug-resistant bacteria will develop, which creates a problem for everyone. From an economic perspective, the societal costs of improper use completely outweigh any private or public benefit.

Because these costs are suffered by third parties who have no say in the transaction, the people prescribing or taking the drugs do not have any incentive to take them into account. We live in an increasingly connected global world, with all the benefits that that brings, but it also means that bacteria can travel easily. Therefore, when considering how to tackle the problem of drug resistance and overuse of antibiotics, we have to maintain a global perspective. A problem in one part of the world can quickly become a problem for all of us.

Raising Awareness and Changing Behavior

In Chapter 5, we saw how raising awareness and encouraging behavioral changes could help prevent infections, through better handwashing, for example. Such techniques could also help to reduce the demand for unnecessary antibiotics.

One reason people take too many antibiotics is that they are simply unaware of the problem of antimicrobial resistance. The Wellcome Trust did interviews and convened focus groups in London, Manchester, and Birmingham to assess public knowledge of antibiotic resistance. Most people, even if they had heard of the problem, thought that taking too many antibiotics could cause their own body to become resistant. They did not understand that it was the bacteria that develop resistance. This is an easy mistake to make, but it means that people often do not realize that their use of antibiotics can have an effect on other people, and conversely, that other people's misuse could also have a negative effect on them. "Few of those interviewed," reported the Wellcome Trust, "think they overuse or misuse antibiotics, so therefore they mistakenly think resistance will not be a problem for them."

This lack of understanding can also result in patients demanding antibiotics from their doctor even when they are unnecessary. The Wellcome Trust study noted that many people wanted to receive a prescription for antibiotics because it confirmed—for themselves, their families, and their coworkers—that they were actually ill. A survey by the UK Longitude Prize in 2014, which received responses from over a thousand UK doctors, showed that around 90 percent of them had been pressured by patients to give them antibiotics.

Poor awareness also means that people find it hard to understand how their actions are a small part of a much bigger problem, just as it is hard to see the connection between a short car drive and damage to the environment. When we spoke in November 2016 with the economist Nicholas Stern, chair of the Grantham Research Institute on Climate Change and the Environment, he drew parallels between antimicrobial resistance and climate change: "People find it difficult to see the cause of death, and they find it difficult to find something immediate in an effect that's coming through with a long lag." People have trouble visualizing the consequences of a slow-moving phenomenon like climate change; compared with a terrorist attack or a plane crash, the link between cause and effect is much less dramatic. In addition to having trouble connecting cause and effect, we underestimate the scale of the threat from both antimicrobial resistance and climate change. "We have to find ways of getting people to understand that this is as bad as terrorism but on a far, far, bigger scale," said Stern. "Just a completely different order—you're comparing just a few thousand with many millions."

There are ways to improve awareness of these issues. Campaigns to help people better understand drug resistance and its impact could make them less likely to pressure physicians into giving them an antibiotic. Patient-advocacy organizations focused on HIV, diabetes, cancer, and elderly care have a vitally important role to play because patients with these conditions are particularly susceptible to infection. Many of these groups have strong fund-raising bases and influencing power that could be used to good effect.

Another way to influence patients and prescribers is by setting up systems that "nudge" them to make better choices, using the principles of behavioral economics pioneered by psychologists Daniel Kahneman and Amos Tversky and popularized in the book *Nudge*, by Cass Sunstein and Richard Thaler. A study in the United Kingdom by Michael Hollingsworth and colleagues explored whether the insights from behavioral economics could be used to change the prescribing behavior of doctors. The researchers selected the general practitioners (GPs) who were in the top 20 percent in terms of the number of antibiotics they prescribed in their region. These GPs were then divided into two groups, with several hundred practitioners in each group. One group received a letter from Sally Davies, chief medical officer for England, which briefly mentioned the threat of antibiotic resistance and informed each GP that 80 percent of the other physicians in their area prescribed fewer antibiotics than they did. The result? Over a six-month period, the prescription rate was 3.5 percent less among the group that received the letters than among the group that did not. Davies describes the result as "quite impressive—given it was just a letter." Not only is it impressive, it is also an inexpensive, quick, and easy way to make an impact. The study concluded that the letter resulted in 73,406 fewer antibiotic prescriptions.

Another campaign that shows promise for reducing unnecessary use of antibiotics is the Red Line Campaign, which was started in 2016 by the health ministry in India, where rates of antibiotic use are high. Pharmaceutical companies were required to print a thin red line on all packages of antibiotics. The idea was to remind people that antibiotics are different from other medicines through a simple visual indication. This initiative could be a starting point for the labeling of antibiotics internationally. The effect of interventions such as these should be quantified so that we can determine which ones are most effective in decreasing prescriptions for antibiotics.

Although they have the potential to make a real difference, these measures can only go so far in tackling the problem of overprescription as long as physicians do not have more accurate ways to diagnose. If doctors simply write fewer prescriptions, it is inevitable that some patients might not receive an antibiotic when they really need one, potentially leading to serious health effects. Such unintended consequences of antibiotic stewardship are no doubt in the back of many doctors' minds when they are prescribing. In discussions we have had with experienced physicians, most recount an incident of having decided not to give an antibiotic to someone who as a result suffered with an infection for longer than was necessary, or even died. It is understandable that these experiences shape physicians' decisions. Although some reduction in unnecessary use can be achieved through training and by encouraging better behavior, such initiatives are not enough. Accuracy in diagnosis is what we really need.

Improving Diagnosis

A significant proportion of doctor visits are for respiratory and flu-like symptoms. Since empirical diagnosis is not exact, some patients will receive antibiotics when they do not need them, and some will not receive them when they do need them. In addition, these visits take up a great deal of time, adding to pressure on budgets and leaving less time for doctors to spend on other patients.

Improving diagnostic techniques could help to reduce antibiotic use as well as saving physicians' time. One proposal is for pharmacies to play a larger role in handling certain types of relatively low-risk infections, such as strep throat. In 2016, England's National Health Service (NHS) announced that it would be rolling out a sore throat "test and treat" service in pharmacies. Patients would be able to go to a community pharmacy for an instant screening test for strep throat, thus relieving the pressure on doctors, who currently see patients with these symptoms 1.2 million times per year. In the future, doctors' offices may be equipped with more complex rapid diagnostic devices that would be able to indicate the type of bacterium causing the infection, whether it is resistant, and which antibiotics might be used to treat it. These kinds of technologies should reduce the number of unnecessary prescriptions, thereby lowering the rate of generation of drug-resistant bacteria.

Given the central place that technology holds in our lives, it is astonishing that technology companies have not put more resources into fixing this global problem. Advanced computer systems and artificial intelligence (AI) could play a much bigger role in shaping diagnosis and prescription. While the up-front costs of using such technology may be sizeable, the long-term benefits to the health-care system need to be factored into value assessments.

We believe that AI platforms could improve on the empirical prescription approach. Physicians work long hours under stressful conditions and have to keep up to date on the latest medical research. To make this work more manageable, the health-care system encourages doctors to specialize. However, the vast majority of antibiotics are prescribed either by generalists (e.g., general practitioners or emergency physicians) or by specialists in fields other than infectious disease, largely because of the need to treat infections quickly. An AI system can process far more information than a single human, and, even more important, it can remember everything with perfect accuracy. Such a system could theoretically enable a generalist doctor to be as effective as, or even superior to, a specialist at prescribing. The system would guide doctors and patients to different treatment options, assigning each a probability of success based on real-world data. The physician could then consider which treatment was most appropriate.

Apps and specialized websites could also be helpful during consultations. If a doctor could see what infections have been spreading in a geographic area and which treatments have proven effective, the information would assist in determining how to treat a patient. Such information could also be presented in visual form that doctors could refer to when explaining their diagnosis. Imperial College London has developed an online tool that uses Public Health England guidelines to help physicians make better decisions when prescribing antibiotics. Called the

Point of Care Antimicrobial Stewardship Tool (POCAST), it enables physicians to navigate large amounts of information and can also be used by members of the public, thereby raising awareness about appropriate antibiotic use and drug resistance.

Improving Dosing

Most studies on dosing of antibiotics were undertaken in the 1970s, when there was not much awareness about drug resistance and people on average weighed less (and thus required lower doses of medication). As a result, we often do not know what the optimum dose is for a patient with an infection, and doctors may prescribe treatment courses that are longer than they need to be, or that call for too low a dose. There have not been a sufficient number of studies carried out to give doctors the information they need to prescribe antibiotics in the most effective way.

Such studies would require large clinical trials, and running clinical trials is very expensive. Pharmaceutical companies have little incentive to conduct studies on proper dosing of antibiotics because the results are unlikely to increase their sales. Governments and other international funders need to step in to encourage more research.

Even in cases where dosing is well understood, however, that advice is often not heeded. For example, it is well established in the medical literature that a very short course (one or possibly two doses over less than twenty-four hours) of pre-operative antibiotics is the correct duration for most "clean" surgical procedures (where there is little contact with contaminated material), such as hip replacements. There is no benefit to the patient in having a higher level of antibiotics in the bloodstream before surgery, and if the wound is dressed properly afterward, the risk of postoperative infection is low. Nevertheless, a survey conducted in 2011–2012 by the European Centre for Disease Prevention and Control found that in 70 percent of the participating countries, more than half of surgical procedures were preceded by prophylactic antibiotic courses lasting more than twenty-four hours.

The survey also found quite a lot of variation among countries, with some offering antibiotics for up to five days before surgery. At least part of this variation may be attributable to cultural factors, in particular, the extent to which a particular society tolerates uncertainty. A doctor might prescribe a higher dose just to be on the safe side or to "go the extra mile" for their patient. In addition, pressure from the patient and the patient's family often pushes doctors into giving longer antibiotic courses than are required. We need to improve the information that is available to doctors and train them to change their behavior.

Although it is unfortunate that doctors do not have better information on dosing, they are still by far the best placed to diagnose infections and prescribe antibiotics. Nothing in this chapter should be read as justification for finishing a prescription early or ignoring a doctor's advice—that advice is invariably far better than a patient's best guess.

Improving Disease Surveillance and Data Collection

Surveillance is a crucially important part of addressing the problem of overuse of antibiotics: if we cannot track the problem effectively, it is very hard to get a handle on it. Currently, we do not collect enough data on infectious diseases or antibiotic use globally. In many countries, information is lacking about which drug-resistant bacteria are circulating. This is partly because of limited testing, and partly because of a lack of the laboratory infrastructure required to process tests and deliver accurate results. To help remedy this situation, the UK government announced in 2015 that it would create a fund called the Fleming Fund (after Alexander Fleming, the discoverer of penicillin, as discussed in Chapter 1) to build laboratories and collect surveillance data in low- and middle-income countries. The fund committed £265 million ($339 million) over a period of five years. The effort will provide more information on drug resistance in funded areas, improving the understanding of the problems locally, nationally, and internationally, and enabling governments and health-care systems to respond.

Further international support is needed, including improved coordination between private and public sectors. Many pharmaceutical companies maintain in-house data on drug resistance rates, which should be made accessible to health-care professionals and policymakers.

We also need to improve the quality and timeliness of data collected. At present, it is often out of date by the time it is released for global disease mapping and surveillance. This is a particular problem for infectious diseases and drug resistance, since infections can spread quickly. Organizations such as HealthMap and the Centre for Disease Dynamics, Economics and Policy (CDDEP) have already developed easy-to-use online world maps that offer visual displays of reported infections. This technology vastly accelerates the pace of disease reporting and has already improved the way experts monitor resistance. It has the potential to be revolutionary, if organizations are given access to more real-time data. The Wellcome Trust and the UK Department of Health are both funding a project coordinated by the Institute for Health Metrics and Evaluation, an independent research center at the University of Washington, to determine how we can make better use of the data that are already being collected. The institute is working to monitor the progress and global burden of drug resistance using innovative data collection techniques and robust modeling.

Rapid diagnostics devices also have the potential to play an important role in collecting large amounts of new data. Such devices are likely to find a much more prominent place in most health-care systems in the future—although when that occurs will depend on the ambition of governments, health-care systems, and international funders. In theory, we could conduct millions of drug-resistant infection tests annually, or even monthly, and enter the data into databases for analysis. Data would be uploaded to the cloud, anonymized, and available for use instantaneously. Our ability to take decisive action would no longer be hindered by a time lag.

In addition to improving data collection, it is vital to establish structures and agreements to ensure that the data are used in a way that will benefit global public health while maintaining accessibility and security. We need to formulate agreements on ownership of the data and how it will be processed and shared, which will require the coordinated efforts of the diagnostics industry and surveillance experts, as well as governments and health-care systems. We recognize that governments, health-care systems, and data providers may have concerns about sharing large amounts of health data internationally, for personal privacy and economic reasons. Release of detailed data on the prevalence of drug-resistant infections could affect tourism, especially health tourism, which brings large numbers of patients to certain countries because their health-care systems have a reputation for high quality and low cost. There are also understandable concerns about personal privacy when sharing any form of health data. And finally, a private company might refuse to collect or share data if not reimbursed. These are all difficult questions that can and should be solved quickly.

Obstacles to Developing Rapid Diagnostic Tests

Why have we not changed the way we diagnose most infections since antibiotics came into use? There are two core reasons: the scientific and technical challenges, and the lack of economic incentives.

John Rex, the expert on antimicrobial resistance whom we met at the beginning of the chapter, spoke to us about some of the fundamental scientific challenges. It is more difficult to prove that certain bacteria are causing an illness than others. "Tuberculosis is never a normal pathogen [in the human body]; any little signal of tuberculosis shouldn't be there," he explained; in contrast, "for the most common infections—urinary tract infections, skin infections—the bacteria that cause them are also normally present in and on the human body." Much of the time such bacteria are harmless, unless we cut ourselves or suffer from an injury or illness that compromises our immune system. Thus if a test detects one of these bacteria, it is difficult to know whether it is the root of the problem or just something that happens to be present but is causing no problems.

A number of technical hurdles must also be overcome before rapid diagnostic tests will be useful. These tests must be fast, accurate, and easy to use. We already have lab tests that can confirm whether an infection is bacterial or viral, which bacteria are involved, and whether they are susceptible to the antibiotics that are available, but these tests take at least thirty-six hours. By that time, treatment will already have started. Accuracy is especially important in low-income or remote areas. A doctor in a high-income country might be willing to accept a less accurate test, because if the test fails to detect a bacterial infection, she can always advise the patient to return if symptoms worsen. However, the doctor might not want to take that risk in a location where a patient would be unable to return for a reassessment. Similarly, in an acute setting, a doctor faced with a patient with suspected sepsis—an infection that causes the body's immune system to go into overdrive and has a very high fatality rate—would be unwilling to accept the risk of not treating the patient with a broad-spectrum antibiotic. Finally, ease of use is important particularly in community settings, such as pharmacies or rural clinics, where staff may have little or no formal training.

Despite these scientific and technical challenges, there are companies that have products that are very close to market and could jump some of these hurdles. Unfortunately, major economic and regulatory challenges create obstacles that have prevented these products from being developed further and adopted. Companies often have to provide evidence of the effectiveness of their product in different countries, regions, or sometimes even hospitals within the same country. These regulations increase the burden on companies trying to roll out a rapid diagnostic test, decrease the rate at which innovative products can be manufactured in high volumes, and have financial costs for developers.

There is also an economic problem that runs to the heart of why there has been such slow progress in developing rapid diagnostic tests. The cost of running a test is typically paid for by the individual patient or physician. Since most antibiotics are relatively cheap, this cost is likely to be higher than the cost of the office visit and the prescription. It is usually cheaper to use empirical prescription, with its risk of overprescription, than to use the diagnostic test. Comparison of these direct costs is, as we have seen, misleading, because it does not take into account the fact that regular use of diagnostics would reduce unnecessary antibiotic use, benefiting society and the broader health-care system.

These costs and benefits can be particularly difficult to calculate because they often accrue to different areas of the health-care system. For example, a physician with low, accurate rates of antibiotic prescription as a result of using a rapid diagnostic test would probably save costs for the local hospital, since fewer drug-resistant bacteria would be circulating in the community, leading to fewer patients admitted to the hospital with drug-resistant infections, which are very expensive to treat. (A study by Tufts University estimated that treating a drug-resistant infection costs between $18,588 and $29,069 per patient.) Additional diagnostic tests in the hospital would not only determine the right treatment more quickly, but would allow hospital staff to separate patients with dangerous infections from other patients and reduce the chance of a wider outbreak, resulting in hard-to-quantify benefits for the community as a whole.

Jonathan O'Halloran, of the diagnostics company QuantuMDx, points out that not enough studies have been done on the economics of rapid diagnostic products to assess their benefits or their potential impact in health care. "No one has done the health economic studies, so no one knows what the true value of an accurate diagnostic test is at this moment. . . . Small diagnostic companies have a very limited pool of money, and they have to be focused to get the product to market." The industry has little ability to build an evidence base. There are many small developers, but they do not have the funds, and sometimes also lack the expertise, to do these studies. This problem is compounded by the fact that research on the benefits of a diagnostic would probably be applicable to other similar products as well, thus benefiting competing companies as well as the company funding the study, and reducing the incentive for any one company to do the research. Competition among companies also makes collaboration unlikely.

Chantal Morel, an expert in diagnostics at the London School of Economics and Political Science, believes that there has been too much focus on the narrow, short-term cost-effectiveness of new diagnostics. When we spoke with her in November 2016, she remarked that people in the public sector "have not taken a sufficiently long-term perspective. They have not looked at the impact of misusing antibiotics over time and the cost of having to treat resistant infections that arise from this misuse (if indeed we are even able to). Appropriately designed . . . studies are urgently needed to demonstrate the cost-effectiveness of utilizing new diagnostics."

We believe that more health economics studies need to be done to give a clearer indication of the value of diagnostics. Health-care systems and nongovernmental organizations (NGOs) should help fund these studies, which cost more than any single diagnostics company could bear, but industry also needs to step up by partnering with these organizations and providing co-funding. All of these groups have incentives to make this happen—whether it be increased sales, or lower rates of antimicrobial resistance as a result of more accurate prescribing.

Encouraging Innovation

In addition to research into the large-scale benefits of rapid diagnostics, we can directly encourage both the development of new products and the use of products that are already available, or soon will be.

First, to encourage more early-stage research, rapid diagnostics should be one of the key beneficiaries of increased innovation funding, as discussed in Chapter 4. Early-stage funding is not enough, however; developers of diagnostics need to know that there is a market for their products. We propose that governments and health-care systems in high-income countries mandate the use of rapid diagnostics and provide funds for devices that have been shown to be cost-effective and clinically beneficial. A deadline for mandatory use should also be set. Such a deadline would signal that health-care systems see rapid diagnostics as a priority and are committed to purchasing successful products.

Setting the deadline further in the future, as long as there was little chance of the policy being abandoned, could stimulate earlier stages of the development process. Products in early development would then have a much more certain market and prospect of future sales. New venture capital could be unleashed, since early-stage products would have more commercial promise. Additional developers might also become interested. Measures would also be required to promote competition, thus ensuring fair pricing. Governments and health-care systems could pilot such an approach. As an example, they could mandate the use of a rapid diagnostic test before an antibiotic could be given for strep throat. This requirement would lead to increased testing and reduced use of unnecessary antibiotics. The marketplace would encourage companies to compete to produce better tests. The lessons learned from such an exercise could provide an impetus for mandatory use of diagnostics on a broader scale in high-income countries. Although an international agreement to require testing would be ideal, even a bilateral agreement between a couple of high-income countries could have a significant effect on the development of new rapid diagnostics. Under this proposal, patient safety would remain the top priority. If the stimulus did not provide a suitable and effective test by the deadline, the mandatory use of tests would not be enforced, and governments would need to reassess whether additional interventions were needed to support development.

These proposals could provide solutions for high-income countries, but many countries would not be able to afford widespread use of rapid diagnostics. Because drug resistance is a global problem, high-income countries, NGOs, and development organizations such as the World Bank have an incentive to provide support to low-income countries to increase use of rapid diagnostics. We believe that a model similar to the one used by a group called Gavi, the Vaccine Alliance, could be very effective. Gavi is a public-private partnership whose mission is to increase access to vaccines in poor countries. To raise rates of vaccination against a preventable bacterial infection called pneumococcal disease, Gavi developed the pneumococcal advanced market commitment (AMC). The Gavi pneumococcal AMC used donor money to provide a market price for vaccines once they had been developed, so companies knew that if they invested money developing a new vaccine, Gavi would buy it. Developers signed legally binding commitments to ensure that vaccines were available at prices affordable to low-income countries. Providing such financial assurances achieves a few different objectives. First, it promotes vaccine research and development, since companies know that they will receive a return on their investment. Second, the model increases rates of vaccination by providing more predictable prices to government funders.

In our version of this model, which the Review on AMR called a "diagnostic market stimulus," additional payments would be made to the developer once a diagnostic product was purchased, in order to partially subsidize it. This model would provide incentives for companies to create diagnostics that could work in low-income settings. To encourage testing, the amount of the subsidy could be determined by how many tests had been performed. This flexible approach would suit this particular market, since different types of diagnostics are useful in different settings. For example, one diagnostic might be very fast, another might be particularly accurate, and a third might be the simplest to use. Each one of these would be the favored choice in different circumstances. Ideally, financial incentives would be set up to reward all types of innovations for the diverse interests and needs of purchasers.

We believe that such an incentive would represent excellent value for the money. The Review on Antimicrobial Resistance forecast that spending $1 to $2 billion per year would have a significant impact on efforts to increase use of rapid diagnostics and vaccines around the world. Given the immense costs of antimicrobial resistance, which dwarf this figure, the case for action is clear.

In this chapter we have considered a range of areas that need to be improved in order to reduce the unnecessary use of antibiotics in humans. These include doing further dosing studies to make prescriptions more accurate, undertaking better surveillance of infectious disease, and changing the behavior of doctors and patients. Finally, we outlined how rapid diagnostics should be central to these efforts. Such diagnostic tests are on the horizon, but it is essential to accelerate their development: the longer we wait, the more pressure we place on existing antibiotics. Without the right economic incentives, they will continue to be only a promising area of interest. With the right incentives, they could transform the way infections are diagnosed and treated, and provide a strong step forward in the war against drug-resistant bacteria.

Seven

Agriculture and the Environment

Guy Poppy, chief scientific adviser at the UK's Food Standards Agency (FSA), works at the interface of food and agriculture. When we interviewed him in November 2016, he recalled first grasping the enormous extent of the drug resistance problem: "I realized the tremendous effort being utilized in hospitals and, especially, general practices [to combat drug-resistant infections], that were not being replicated on farms. . . . And yet the evolutionary selection pressures are still present, and, in fact, are probably more likely to create resistance. In agriculture, animal welfare is compromised and food prices escalate as intensive farming methods become more widespread. Other methods are needed to rear animals, with no antibiotics in the toolbox. The problem is as real as climate change, already playing out in front of our eyes."

Poppy described an FSA study that tested for bacteria in chicken being sold in grocery stores. High levels of resistant *Campylobacter* were detected, "which means millions of chickens in UK supermarkets contain drug-resistant bacteria." Even though the bacteria are killed by thorough cooking, they can still spread if the meat is not cooked or handled properly. Poppy believes that consumers have the potential to be a very powerful force for change, but progress on an international scale will be a challenge because many meat-producing countries have a financial interest to continue antibiotic use.

Individuals who work closely with livestock, such as veterinarians and farmers, have an important role to play in helping combat antimicrobial resistance.[1] We spoke with the chief veterinary officer of Australia, Mark Schipp, in May 2017. Schipp has both personal and professional interests in this issue—his father passed away from a MRSA infection. The main adviser to the Australian government on animal health issues, Schipp believes that "the industry seems to be a bit complacent," yet he notes that "some of the large retailers and food processors are requiring animals to be raised without antibiotics." This kind of pressure is crucial for changing farmers' behavior and the world's current reliance on antibiotics in livestock rearing. "The only avenue that I can see is a commercial one," says Schipp. "Australian farmers, because we are exporting 65 to 70 percent of our production, are very sensitive to overseas markets, and very responsive to those [pressures]."

In previous chapters, we have mostly been concerned with unnecessary antibiotic use in humans. In this chapter, we examine the broader use of antibiotics in agriculture and aquaculture, as well as the threats posed by antibiotics entering the environment through animal and human waste and manufacturing. We discuss the steps that have already been taken to address these issues and propose solutions for taking further action.

Antibiotic Use in Livestock and Fish Rearing

The extent of antibiotic use in animals, particularly those that are reared for food production, may be a surprise to many readers. Globally, more antibiotics are used for animals than for humans, according to most estimates. Over 70 percent of medically important antibiotics in the United States, by volume, are sold for use in farm animals.[2] This number is partly a result of the sheer number of animals being reared and slaughtered every year to feed the world's seven billion-plus people. It also results from the fact that antibiotics are used not only to treat infections but also to prevent them and to promote growth.

Antibiotics began to be used widely in agriculture after they were first mass-produced for human use in the 1950s. Rates of use increased especially after it was discovered that giving regular low doses of antibiotics to farm animals made them grow faster and larger. In a 2016 article, Jeremy Farrar, director of the Wellcome Trust, described how this discovery came about:

> Like so many breakthroughs, it happened almost by accident. In the late 1940s, fishermen near Lederle Laboratories, in New York state, noticed that the trout they were catching were getting bigger. Word reached a biochemist called Thomas Jukes, who thought it might have something to do with the run-off from Lederle's latest miracle product—an antibiotic called aureomycin. So he and his colleague Robert Stokstad tried an experiment. They took some newborn chicks, and fed one group on a liver extract, designed to cure anaemia. Another was given aureomycin. The results were startling: the birds given the antibiotic did not just survive, but put on weight with extraordinary rapidity. When this discovery was officially announced in 1950, the *New York Times* proclaimed that aureomycin's "hitherto unsuspected nutritional powers" would have "enormous long-range significance for the survival of the human race."

Because of this effect on growth, antibiotics began to be used in the United States and around the world, increasing meat production and lowering prices for consumers. According to an article by the historian Maureen Ogle, "Farmers wasted no time abandoning expensive animal proteins in favor of . . . infinitesimal, inexpensive doses of antibiotics. Their livestock reached market weight more quickly, and farmers' production costs dropped. Consumers enjoyed lower prices for pork and poultry." Major food producers quickly embedded routine antibiotic use into their production systems in efforts to promote growth, in addition to using them to treat animals when they became sick. Interestingly, nobody really knew exactly why the antibiotics had such an effect on growth. Indeed, we are not certain even now. One hypothesis is that they alter the animal's microbiome—the naturally occurring balance of bacteria in the gut.

Modern farming quickly became dependent on antibiotics. Unfortunately, as Neil Woodford, head of the antimicrobial resistance unit at Public Health England, explains, "any use of antibiotics in any sector will select for resistant bacteria." Drug-resistant bacteria can be transferred to other animals, to people (through contact with farmers, in slaughterhouses, or through the incorrect handling or cooking of meat at retail outlets or in the home), and to the wider environment (through leaching of antibiotics into the soil and water systems after excretion). Overuse of antibiotics in agriculture, just as in humans, creates negative externalities.

This situation is clearly a problem for animal health, but it also threatens food security because it encourages the development of untreatable infections. In China, 500 million pigs are processed for food every year—a number which is likely to increase as the middle-class grows. Should a drug-resistant infection wipe out even 10 percent of these stocks, China's ability to meet the demand for pork would be significantly reduced. In addition, producers would be unable to treat the sick animals, and the bacteria could potentially spread to humans.

Risks to Human Health

It is fair to wonder why more has not been done to reduce antibiotic use in agriculture, especially since many reports have called for action, including a well-known report published in the United Kingdom in 1969 that recommended banning the use of human antibiotics as growth promoters for animals. One of the reasons for the slow progress is that there has been a protracted debate over whether giving antibiotics to animals can in fact cause problems for human health, and if so how much this contributes to overall levels of drug resistance. We argue here that the evidence shows that there is a clear risk to humans, even though it has not yet been possible to accurately measure how much of the total drug resistance problem is caused by use of antibiotics in animals.

Unfortunately, the antibiotics we use in livestock are very similar to the ones we use in humans. We know that bacteria can travel from animals to humans in a number of ways, and so if bacteria in animals become resistant to the same drugs that are used for humans, that raises grave concern about the spread of drug resistance to humans. Of the forty-one antibiotics approved by the FDA for use in food-producing animals, thirty-one are deemed to be medically important for humans. Note that we have not picked out the United States for this example because it is the worst performer—far from it. Experts believe that the problem is much worse in many other countries, but the relevant data are not available. However, this example indicates how interlinked human and animal health systems are. Even in a high-income country with a relatively strong regulatory system, drugs relied on to treat a sick hospital patient are the same ones that are used in our farming and aquaculture systems.

In late 2015, in China, researchers discovered a gene in bacteria that was resistant to the antibiotic colistin and could be transferred to other bacteria via plasmids. Colistin, as we discussed in Chapter 2, is a last-line antibiotic for humans, but it is also used extensively in livestock in some countries, including some in Europe. This discovery caused great concern. In an interview in November 2016, Margaret Chan, who was then director general of the World Health Organization, stated, "If we lose colistin, as several experts are predicting, we lose our last medicine for fighting a number of serious infections."

Indeed, a report in February 2017, less than two years after the original paper was published, showed that the gene had been found in over thirty countries on five continents, suggesting that it had either spread quickly or had existed undetected for some time. To guard against the latter scenario, Neil Woodford emphasized how important it is to have effective surveillance for emerging threats: "[The resistant gene] has been there for a while, but it just went unrecognized. There are probably many types of resistance out there that we do not yet recognize. Maybe they have not got into a bacterium that's caused an infection. It has the potential to be very worrying. For the countries that have colistin available as a clinical option, it is usually the last resort."

Unfortunately, by the time colistin had been recognized as the last line of defense against certain multidrug-resistant bacterial infections, it was already being used extensively in some farming systems. The study on colistin resistance in China, for example, found the transferable resistance gene in bacteria from about 20 percent of animals tested. The researchers also found the same gene in approximately one percent of humans tested. (This study pre-dated colistin being available for human clinical use in China, which happened in 2017.) This finding suggests that the resistant gene first appeared in the animal sector and then began to transfer to humans. When we interviewed her, Margaret Chan stated that "the Chinese study . . . connected all the dots, as the gene was detected in samples from farm animals fed colistin as a growth promoter, [from] chicken meat and pork, and [from] humans." Once it became clear that colistin was a critical drug for humans, more should have been done to ensure that it was either restricted or banned from use in agriculture.

So far, it has not been possible to accurately determine the extent to which antibiotic use in animals leads to the development of drug-resistant infections in humans, because it is difficult to track the origin of a bacterium. If a person develops a urinary tract infection in the hospital, for instance, it might be possible to determine what kind of bacteria are responsible (usually *E. coli*), but it is very difficult to establish with certainty where those bacteria came from and when they developed resistance. They could have come from livestock, from a person, or even from somewhere in the environment.

Even so, there is good evidence that agricultural use is one of the factors in increasing drug resistance. The Review on Antimicrobial Resistance conducted a literature review analyzing 192 papers that had tried to answer the question of whether antibiotic use in agriculture has led to drug resistance in humans (see Figure 7.1). Of these papers, 114 found evidence linking animal consumption of antibiotics to drug resistance in humans, 63 did not establish a link in either direction, and 15 presented evidence showing no link. The vast majority (88 percent) of the papers that found evidence of a link were written by independent academics, whereas almost half of the papers that found no link were written by people in either industry or government. Given the amount of research indicating that use of antibiotics in agriculture can lead to drug resistance in humans, we believe that action must be taken now.

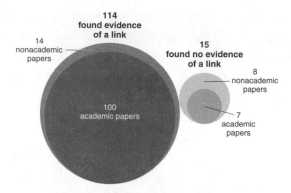

Fig. 7.1. Most published papers provide evidence linking animal consumption of antibiotics to drug resistance in humans. The proportion documenting such evidence is much larger for papers written by academics (100 vs. 7) than for papers written by government or industry researchers (14 vs. 8). Redrawn from Review on Antimicrobial Resistance (CC BY 4.0).

The Economics of Agricultural Antibiotics

In addition to uncertainty about cause and effect, another factor delaying reductions in antibiotic use in agriculture is the economic impact—which is hotly debated. Although the economic effects will vary depending on the type of farm, climate, and many other factors, the use of antibiotics for growth promotion is driven purely by economic considerations, and it is therefore essential to determine what effect a reduction in use would have on farming systems.

The first thing to note is that the economic benefit of using antibiotics to promote growth, at least in high-income countries, seems to have been decreasing over time. An important piece of analysis by Laxminaryan, Van Boeckel, and Teillant in 2015 showed that studies conducted after 2000 in the United States, Denmark, and Sweden have shown much lower benefits from using antibiotic growth promoters than those conducted before 1980. In the newer studies, growth promoters led to an increase of less than one percent in growth in most cases, except for nursery pigs, where an increase of 5 percent has been reported. (Growth increases of around 6 to 12 percent had been reported in earlier decades.) While any economic benefit is desirable to the farmer, a lower growth gain should reduce the incentive to use antibiotics. What caused this change? A leading theory is that antibiotics are more effective growth promoters when used for animals kept in cramped, dirty, unregulated conditions than for animals living in cleaner, more open, more controlled environments. Under suboptimal conditions, the growth promoters are for all practical purposes a substitute for good infection prevention and control. New breeds and better nutrition can also have an impact. As farming systems improve, the benefit of using antibiotics for growth promotion tends to fall, and in high-income countries this benefit now seems to be very small or negligible. Countries such as Denmark and the Netherlands have significantly lowered levels of use, yet they remain highly productive and competitive. Although Denmark uses substantially lower levels of antibiotics for livestock rearing than most other high-income countries, pork productivity has gone up, and it is one of the largest exporters of pork in the world. The country accomplished this through regulations to limit use, along with improved infection control procedures—which lowered infection rates and reduced the need for antibiotics for disease control. Denmark also improved the monitoring of antibiotic sales and use, which enabled the government to intervene if farmers were still overusing antibiotics. It did this through what was called a "yellow card system"—those pig farmers using the most antibiotics were sent warnings that they might face penalties.

The potential economic impact of reducing the use of antibiotics as growth promoters in low- and middle-income countries is less well researched and is likely to vary by country, climate, species, and many other factors. Detailed regional and country-specific studies on economic impact would help to guide policy action at the country or regional level, and encourage reduction in unnecessary use.

We also need to acknowledge that considering the basic economic benefit of growth promoters oversimplifies the problem. We know that antibiotic use encourages drug resistance. Therefore, continued overuse of antibiotics poses a risk of increased drug-resistant infections in animals, which comes with an economic cost. The World Bank produced a detailed report on the economic costs of drug-resistant infections in September 2016. The report modeled how higher rates of antimicrobial resistance would affect livestock production as a result of decreased productivity (due to more untreatable disease) and reduced exports (due to restrictions imposed by trading partners). The burden of these costs, it concluded, would likely fall disproportionately on low- and middle-income countries, for whom agricultural production is a larger share of their GDP.

A Growing Consensus for Action

The World Health Organization (WHO), the Food and Agriculture Organization of the UN (FAO), and the World Organisation for Animal Health (OIE) have all recognized the menace of antimicrobial resistance. In 2016, a high-level meeting on the topic at the United Nations resulted in an agreement that was signed by all 193 countries. This document stated that the overuse and misuse of antibiotics in agriculture is one of the drivers of resistance. "[Antimicrobial resistance] is a problem not just in our hospitals, but on our farms and in our food, too. Agriculture must shoulder its share of responsibility, both by using antimicrobials more responsibly and by cutting down on the need to use them, through good farm hygiene," said José Graziano da Silva, director general of FAO, commenting on the UN meeting.

Another international organization that has expressed concern is the G20 (a group of countries comprising twenty of the world's largest advanced economies). In an official statement issued in 2017, the group announced: "We will promote the prudent use of antibiotics in all sectors and strive to restrict their use in veterinary medicine to therapeutic uses alone. Responsible and prudent use of antibiotics in food producing animals does not include the use for growth promotion in the absence of risk analysis."

Although the growing recognition of the issue and commitments to take immediate action are very positive steps, the pace of international action is still not fast enough. Far more attention and resources need to be devoted to the use of antibiotics in the world's farming systems. If not, the danger is that business (with unnecessary antibiotic use) will continue as usual.

Recommended Policy Changes

The solutions to this problem are neither quick nor easy. Kofi Annan, former secretary general of the United Nations, once said, "On climate change, we often do not fully appreciate that it *is* a problem. We think it is a problem waiting to happen." The same is true for antibiotic use in agriculture. It is a problem now, even if we cannot quantify its exact size. Addressing it will raise broader issues about global food production, including how we produce food in an efficient and sustainable way over the long term.

It is important to realize that resistant bacteria already exist in each country and are transferred around the world by international travel and trade. Therefore, just as in the case of human antibiotics, any long-term solution requires a global effort. Yet we also need to be pragmatic. Many of the poorest countries in the world have very low levels of meat production and relatively limited access to antibiotics. They also have fewer resources for putting in place the infrastructure to improve farming practices and training and to ensure appropriate regulatory oversight. They will need support. International development agencies, alongside the OIE and FAO, could play an important role by helping to develop better veterinary systems, farm management systems, and education programs on the responsible use of antibiotics.

While the WHO, FAO, and OIE are correct to encourage action plans from the UN's 193 member countries, in reality, only a small number of countries produce the majority of the world's meat and therefore use most of the antibiotics in livestock. The G20 countries are responsible for around 80 percent of world meat production. For this reason, we believe that the G20 should take the lead on tackling antimicrobial resistance in agricultural settings, building on the strong statement the group released in 2017.

Three interventions have the potential to radically improve the current situation: improving surveillance of antibiotic use and drug resistance in animals, undertaking economic studies of the impact of reduced antibiotic use in low- and middle-income countries, and implementing targets to reduce unnecessary use.

Surveillance of antibiotic use in agriculture must improve. We do not know enough about antibiotic use in humans, but we have even less information on use in livestock and farmed fish. In many countries, little or no data are available. We need to collect data on the amounts and types of antibiotics being used in different animal species, as well as instances of drug resistance. These data must be recorded and shared with governments and regulatory systems. Summary statistics and best practices should be shared internationally.

In 2015, the UK government made a large international funding commitment to help low- and middle-income countries gather data on antibiotic use in both animals and humans by creating the UK Fleming Fund, as discussed in Chapter 6. Sally Davies, chief medical officer for England, who was instrumental in developing this fund, said, "You cannot expect low-income countries to do diagnostics and surveillance as we do in developed countries, unless you support them." Additional funding is required to reach the standard of surveillance needed to monitor and address this problem globally. We urge other countries and international organizations to provide such funding.

As a second step, we need further analysis of the region-specific economic impacts of efforts to reduce antibiotic use. Because there is so much variability in farming systems across the world, it is important to estimate the likely economic cost for different countries. We propose that the World Bank lead this important analysis, since the organization has experience working on drug-resistant infections, as well as on economic and development issues more broadly.

Finally, countries should set targets at the national level to lower the overall quantity of antibiotics given to livestock and fish, while also establishing mechanisms for regular measurement and enforcement. Interventions should be flexible enough for individual countries to set their own goals and determine how they would achieve them, as long as they abide by the broader principles agreed to internationally. Public announcement of each country's target would put pressure on governments to prove that they are meeting their desired levels. A method for enforcing these targets would need to be determined.

There are undoubtedly numerous issues that would have to be worked through to deliver targets. An international group of experts could manage proposals and create a workable program to calculate targets and measure progress. We propose measuring the amount of antibiotic used for fish and livestock in terms of milligrams per kilogram of body weight. The European Medicines Agency requires such measurements of all European Union countries, so there is precedent for this method. We recommend that the group consider setting targets by animal type or species, since their needs differ; fewer antibiotics are required for lambs than for pigs, for example. Other factors, such as climate and disease prevalence, might also affect target values.

A time period for the targets would also need to be calculated. In 2016, the Review on Antimicrobial Resistance proposed targets that countries would begin working toward in 2018, in order to provide time to put surveillance and enforcement systems in place. Targets could then be set for periods of ten years, with milestones to ensure regular progress. However, if governments wanted to be more ambitious, they could, of course, shorten these timelines. The UK government, for example, responding ambitiously to the Review's recommendation, decided to aim for a level of less than 50 mg./kg. (milligram of drug per kilogram of animal biomass) by 2018 (calculated as a national average for all livestock and fish farmed for food). The level at the time of the announcement, according to the most recent figures, was 62 mg./kg. In late 2016, the government announced that it was on track to meet this target. We hope that if targets are made public, it will encourage ambition and positive competition among countries.

Implementing Changes

To implement changes, international agreement on the principle of targets is an important first step. Countries need to consider how they would reach their targets using the levers they have at their disposal. A combination of taxation, regulation, and subsidies for alternatives to antibiotics should be deployed.

Every time a farmer uses an antibiotic on an animal, it imposes a cost to society that is not included in the cost the farmer pays for the antibiotic. This hidden cost is the rising risk of drug resistance. A tax on the antibiotic would "internalize" this cost, raising the price closer to the total cost imposed on society and potentially causing the farmer to purchase less of the drug. Taxes are often used to deal with such negative externalities. To take an example from climate change, drivers have little incentive to think about how their actions affect the environment, since they reap the benefits of driving but pay few of the wider societal costs. These costs are not always apparent to the driver. Local air pollution can be, but the impact of carbon dioxide on the climate is not. If we tax gasoline, the cost to drive will increase, deterring people from driving as much while not preventing those who need to from doing so. The same could be done with antibiotics. If a tax were added that increased the price, a farmer who wanted to use them just to help their animals grow slightly faster or to avoid investing in proper infection control might decide not to buy them, or to buy less. But a farmer who needed the antibiotics to cure a sick cow would likely be willing to pay the higher price. Economists tend to favor taxing negative externalities rather than regulating heavily because it can be difficult to assess whether use of a product is necessary or not. How can we tell if a farmer really "needs" to give an antibiotic to an animal? Changing the price of antibiotics through a tax is a way to directly affect farmers' incentives and decisions. An additional benefit of taxation is increased revenue, which governments could use to counter problems related to antimicrobial resistance, or to help farmers improve infection control facilities in ways that would further reduce the need for antibiotics.

A second way to reduce antibiotic use is through regulation. An example is the 2006 ban on the use of antibiotics for growth promotion in the European Union, which had mixed results. The effect of the ban varies significantly by country: overall levels of antibiotic use fell in countries that implemented additional domestic policies to reduce use, while in other countries, levels are the same today as in 2006. The variance in levels of use across Europe and other high-income countries today suggests that regulation—at least of growth promoters alone—is not sufficient to reduce use across the board (see Figure 7.2).

Why did banning the use of antibiotics for growth promotion not solve the problem? As this category of use formally fell to zero, antibiotic prescription rates increased for other uses,

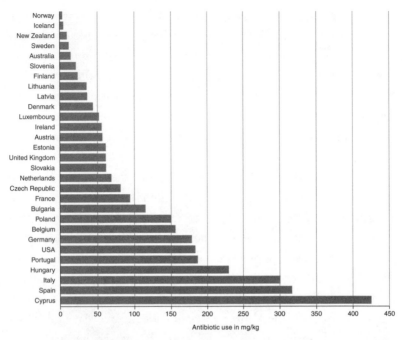

Fig. 7.2. Antibiotic use in agriculture by country, 2011. *Note:* All figures are given in mg. purchased per kg. of livestock biomass and do not include ionophores and oligosaccharides. For New Zealand and USA, animal biomass estimated based on numbers of animals. Redrawn from Review on Antimicrobial Resistance (CC BY 4.0). Data Source: European Medicines Agency (2011) and the national governments of the United States, Australia, and New Zealand.

such as prophylaxis (preventive use) and metaphylaxis (treating a group of animals after one animal falls sick). One of the challenges with this kind of ban is that it is difficult to assess whether an antibiotic is being used for growth promotion. Some farmers might try to "game" the system; others might be relying on growth-promoting antibiotics to help prevent infections as well, and so would want to continue using them for that purpose. For these reasons we do not believe that regulating by type of use is the most effective way to reduce antibiotic use in farm animals and fish.

In addition to regulating the quantity of antibiotics being used, we must limit use of those that are most critical for human health, such as colistin. Restrictions or bans on the use of these critical antibiotics in animals might be effective. Such regulations would not require oversight of the intention of use, and therefore could be easier to enforce than the ban on growth-promoting antibiotics. If the transferable resistance to colistin has taught us anything, it is that we are far too slow to respond. Jeremy Farrar, director of the Wellcome Trust, made this point in an interview with us in November 2016: "We have not got it right yet in emerging epidemics and pandemics . . . but we're . . . moving in the right direction. We do not associate drug resistance in the same way with the need for speed, and yet there is a need for speed, and colistin is a good example. As soon as it became apparent that it was going to be important to reuse colistin in humans, we should have made an instant decision to stop use in the agricultural sector.[3] But it is still being used in industrial doses in the agricultural sector. We need a sense of urgency and pressure which we just do not have." It is understandable that this sense of urgency does not develop as naturally in the case of antimicrobial resistance as for some major infectious disease outbreaks, since its effects are harder to see. Yet if we continue to react without urgency, the large numbers of lives lost to drug resistance will increase every year. While many of these problems are not black or white, there is a strong case to be made that certain types of critical antibiotics should not be used in food production systems at all, and regulation might well be the best lever for governments to achieve this.

A third way to reduce antibiotic use is to promote alternative means to prevent and treat infections so that fewer antibiotics could be used. These alternatives, if subsidized, could play a much bigger role than they do now. They include improving infection prevention and control with better farm management practices and infrastructure, using more robust breeds or varieties, and making better use of vaccines and diagnostics. If antibiotic use was simultaneously taxed or more highly regulated, these alternatives could become more attractive propositions to farmers. Furthermore, providing subsidies for them would accelerate change and allow policymakers to focus on interventions that add more value, such as those that address food security and sustainability. We believe that vaccines, in particular, could be instrumental in reducing antibiotic use on farms and in fisheries by preventing the development of bacterial infections. Norway provides an example of a country that has made vast improvements, dramatically reducing its reliance on antibiotics in fisheries, primarily through the increased use of vaccines. Between 1987 and 2013, antimicrobial use fell by 99 percent, even while the industry grew in size by around twenty times.[4] Once vaccines were developed, strong collaboration between the government, fish farmers, and regulators helped get them into wide use. This progress is extraordinary, and while it may not be able to be replicated in every country in the world owing to different disease challenges, it gives an indication of the scale of the change possible.

Industry's Role in Reducing Antibiotic Use

The food industry can support efforts to reduce antibiotic use, and it has already begun to do so. Providing increased transparency for consumers, including information about how animals were raised and whether antibiotics were used, can inform purchasing decisions. Retailers and fast food outlets could (and should) decide to provide only meat products that meet a certain standard of antibiotic use. Industry has a great deal of power to drive multinational action. In the United States, a number of fast food outlets, food producers, and retailers have announced that they have reduced the use of antibiotics by their suppliers. The Natural Resources Defense Council (NRDC) estimates that around 5 percent of total US meat sales each year are of products raised without antibiotics, and that this share of the market is increasing. It is worth noting that many of these announcements concern the use of antibiotics in chicken. Sales in the United States of "antibiotic-free" chicken increased by 34 percent in 2013, and the NRDC estimates that "more than one third of the entire U.S. chicken industry has now eliminated or pledged to eliminate routine use of medically important antibiotics." This reflects significant progress from the industry, but it is vital that these promises are followed through on. More attention must be given to reducing antibiotic use in the production of pork, beef, and fish.

Encouraging increased transparency from producers by means of an internationally recognized "responsible use" standard could help speed up and broaden progress. Investors also have the opportunity to require better antibiotic use policies among private companies. We spoke in November 2016 with Jeremy Coller, a private equity executive and founder of the Jeremy Coller Foundation, who has investigated how investors can put pressure on food companies. "Investors can use their influence as shareholders to engage with the companies in which they invest to stress the need for both swift practical action and forward-looking innovation," explained Coller. "I work with large, long-term investors, and risks of a systemic nature are of particular concern. The widespread consequences of antimicrobial resistance have the potential not only to destroy the value of individual companies but also to significantly impact economies. Investors are the owners and stewards of food, agriculture, and health-care companies, and it is clear why they are motivated to address this risk." Coller created an enterprise called the Farm Animal Investment Risk and Return (FAIRR) initiative to help investors identify these risks and encourage companies to reduce antibiotic use in their supply chains. "We are already seeing this issue rise up the investment agenda," said Coller. "In 2016 FAIRR brought together a coalition of sixty-one institutional investors, worth $2.2 trillion, to ask ten of the largest US and UK restaurant chains (including McDonald's, Domino's Pizza Group, and Yum! Brands) to take action on the systemic overuse of antibiotics. This has already resulted in companies such as The Restaurant Group announcing that it will phase out the routine, preventative use of antibiotics and refine its use of antibiotics classed as 'critically important' for human medicine. As more investors become aware of the material risks of this issue, I expect antimicrobial resistance to become as much of a mainstream investment issue as corporate governance or climate risk have become."

Given increasing consumer and investor pressure, food production and retail companies need to be proactive in reducing levels of antibiotic use. They should emphasize collaboration and sharing of best practices, following the example of the pharmaceutical industry. In early 2016, one hundred pharmaceutical companies signed a joint declaration, the Davos Declaration, indicating their willingness to support action to tackle drug resistance. This declaration was followed in September 2016 by an action roadmap (the Industry Roadmap for Progress on Combating Antimicrobial Resistance) that called for developing new antibiotics, improving access to vital antimicrobials, reducing unnecessary use in humans and in agriculture, and cleaning up supply chains. Major food producers and retailers should follow this example.

Antibiotics in the Environment

Antibiotics and their residues reach our environment in three main ways. The first two are a result of waste—from both animals and humans. Studies suggest that as much as 75 to 90 percent of

antibiotics may be excreted from animals without being metabolized. This waste goes into the soil and is then washed into water
systems. Although we do not yet have much information on the
impact of this situation on human health, experts agree that it
could promote development of drug resistance. In addition to reducing overall use of antibiotics, we could diminish this pollution
source by improving waste treatment procedures and modifying
the antibiotics so that they are more easily metabolized.

Humans also excrete unmetabolized antibiotics, which then
work their way into water systems. Currently, a wastewater
system that eradicates all traces of antibiotics does not exist,
partly due to the high cost of development. This problem is especially acute for hospital waste, since hospital patients are more
likely to have antibiotic residues in their feces, in addition to
drug-resistant bacteria. This combination has the potential to
create hotspots of resistance. Reduction of unnecessary use
would help mitigate the impact, and hospitals should also improve their treatment of wastewater. The comparatively smaller
cost of hospital wastewater treatment could be a sound investment, especially if wholesale improvement of wastewater management on a larger scale is not economically feasible.

The third way that antibiotics reach our environment is through
the process of manufacturing. This problem is perhaps less frequently discussed, yet it is more easily addressed. During the antibiotic manufacturing process, active pharmaceutical ingredients
(APIs), the essential ingredients that enable antibiotics to work,
are produced. Most manufacturing takes place in India and China,
where production costs are typically lower than in high-income
countries. Some manufacturers release insufficiently treated waste
into water systems near their manufacturing plants in this process.
There are currently no universally accepted standards for API
discharge, which encourages this practice. Without standards,
companies are less likely to take action, particularly since it is

more expensive to treat the waste than to allow potentially dangerous material to be released into the local environment.

Once the APIs have been manufactured, they are sold in large quantities to global pharmaceutical companies, which chemically modify them to create the various antibiotics that are sold to consumers all over the world. Thus we all benefit from cheaper production made possible by the absence of strict standards. But we pay an even steeper cost, which is not well recognized. The highest cost is borne by local communities living near the affected water, which might be used for bathing and washing clothes and dishes. These water systems can become centers for the development of drug-resistant bacteria. But this practice also has more far-reaching consequences, since highly drug-resistant bacteria developed in one location can transfer to a nearby city or region, or even to a country on the other side of the world. As Joakim Larsson of the University of Gothenburg told us in November 2016, "I would argue that if we see it as a risk/benefit issue, then preventing major discharges of antibiotics to the environment is self-evident." Interventions in this area could be very effective, since there are far fewer entry points into the environment from API factories compared with the millions upon millions of entry points of animal and human feces.

Risks to Human Health

Neil Woodford, of Public Health England, spoke about the need for additional scientific study of the risk to human health caused by antibiotic-resistant bacteria and antibiotics being released into the environment: "What we still lack in many cases is a good understanding of the public health risk associated with these non-human reservoirs of resistant bacteria. People have been worried about agricultural sources of resistant bacteria, environmental sources of resistant bacteria, for many decades but we still have very few solid data of the risk that many of those reservoirs pose to humans."

A combination of scientific and financial reasons underlies this lack of research. On the scientific side, the advances in whole-genome sequencing may soon allow us to determine with far more accuracy where drug-resistant bacteria have originated. Such information could help determine where to focus our attentions. On the funding side, we need to take advantage of increased international attention on antimicrobial resistance to unlock further funding opportunities to build the evidence base.

The few studies that have been done on environmental pollution caused by antibiotic manufacturing reveal how alarming the current situation is. A very important study by Swedish researchers in 2007 helped bring this issue to the fore. We interviewed Joakim Larsson, an expert on bacteria in the environment, about this study, which he headed. He told us that the active ingredients of the antibiotics used in many Western health-care systems are often produced by a different company from the one listed on the package, usually based in a different country. Larsson and his team traveled to India, one of the largest producers of antibiotics and APIs, to conduct several studies of a wastewater treatment plant that received effluent from a number of manufacturers of APIs. The levels of APIs being dispersed into the environment were truly shocking—ciprofloxacin, an antibiotic widely used to treat infections in humans, was found in concentrations around one thousand times higher than those needed to kill some types of bacteria. Larsson told us that the concentrations in treated effluent "exceeded those that you would find in the blood of patients taking the medication." His team also found that "these concentrations led to the selection of highly multidrug-resistant bacteria that can also share their multi-resistance plasmids easily with pathogens." Larsson believes this is why it is essential that we halt the release of polluted water from these plants, even though we have not yet been able to establish with accuracy the impact this will have on human health.

The Solutions

As a first step to solving this problem of antibiotic residues being released from manufacturing plants, industry must improve monitoring, and where necessary, clean up supply chains. The pharmaceutical industry has already shown improvement: many companies are increasing transparency and demanding higher standards, but more should be done.

Larsson brought up the textile industry as a good example of how to deal with supply chain issues. The textile industry had been heavily criticized about working conditions on the production lines of many large companies operating in parts of Asia. Larsson believes this criticism drove significant change. "If you buy clothes today from some of the big companies, you can go onto their websites and you can see that this T-shirt is made by this factory here in Bangladesh, in these conditions, and they fulfil these social responsibility codes, etc. The companies are competing with each other on having a good ethical code of conduct. Transparency is needed to do that."

The pharmaceutical industry committed to action with a roadmap that followed up the Davos Declaration in September 2016, in which the industry pledged to "reduce the environmental impact from the production of antibiotics, including a review of the companies' manufacturing and supply chains, and work with stakeholders to establish a common framework for assessing and managing antibiotic discharge." The roadmap forms part of a wider pledge that companies have promised to make good on by 2020 to help address the problem of antimicrobial resistance. However there are no clear targets, which raises the fear that some companies may not make real reforms. Additionally, the roadmap commitments have been signed by a relatively small portion of the whole industry (thirteen companies at the time of writing), which, although significant, means that we need to keep working to convince the remaining companies to join the effort.

We spoke with Lucas Wiarda of DSM Sinochem Pharmaceuticals, one company that is trying to improve standards pertaining to antibiotic waste. He said, "In the absence of science-based discharge standards for antibiotics, and as a first step, we set ourselves an initial target. If needed, we will apply stricter standards." He went on to discuss the situation in the industry as a whole: "The commitments made by the roadmap companies are courageous and promising, and I am hopeful that they will lead to some positive change . . . [but] volume-wise, and thus pollution wise, the majority of the generic bulk industry has not committed themselves. By signing up for the AMR Industry Alliance, established in 2017, they can still do so. Chaired by the IFPMA [International Federation of Pharmaceutical Manufacturers and Associations], the alliance aims to become an industry-wide initiative that will govern and report out progress on the commitments the industry has made in the Davos Declaration and the industry roadmap, including environmental stewardship—I hope that others, including more generic companies, will join the alliance soon and together we will tackle the problem."

Regulation and industry-led action are not mutually exclusive. Even if progress is made by pharmaceutical companies, there is still a strong case for setting reasonable, binding, minimum standards for disposing of APIs. Very few if any such standards exist anywhere in the world. Further analysis to find the appropriate level for these standards is needed, yet experts could set initial levels based on existing evidence.

Another way for regulators to encourage better production standards is to designate a label for antibiotics manufactured according to a high environmental standard. These labels would allow purchasers to make more informed decisions, in the same way that labels allow consumers to purchase antibiotic-free foods. In this instance, the purchaser would often be a health-care professional—perhaps a doctor or a pharmacist—rather than a patient. Labeling products would make it difficult for purchasers to claim ignorance and would put pressure on companies to meet the standard.

As we have seen in this chapter, the problems and solutions to antimicrobial resistance are far wider than just the human use of antibiotics. At their best, antibiotics protect our pets and our livestock when they become sick and need treatment. At their worst, they get into our food supply, pollute our rivers, and promote the development of drug-resistant bacteria that threaten all of us. To protect ourselves and the life-saving function of antibiotics, everyone needs to get involved. First, civil society groups need to have a larger voice and put pressure on policymakers to act. Second, governments need to realize that the problem of antimicrobial resistance is not restricted to one sector. The issue must garner attention from departments of health, finance, agriculture, science, foreign affairs, education, and development. No progress will be made until these departments work together in individual countries and with their international partners. Finally, the private sector needs to follow through on policy proposals, such as those specified in the pharmaceutical industry's roadmap, to reduce unnecessary use. The high level of political attention accorded to this problem, which resulted in international agreements made at the UN and the G20 in 2016 and 2017, gives us hope. But pressure must be maintained, and strong words turned into real action.

Eight

Next Steps

Peter Piot, director of the London School of Hygiene and Tropical Medicine, is known globally for his pioneering work on Ebola and HIV and is well placed to comment on global infectious disease threats. He spoke to us in November 2016 about his first direct experience with drug resistance: "In 1976 I isolated the second only penicillin-resistant [strain of] *Neisseria gonorrhoeae* from Africa. There was an American group in Ghana [who isolated one strain], and then I isolated [a second strain] from a sailor who came from Côte d'Ivoire with gonorrhea. He had what was then an untreatable infection. . . . It kind of threw up in the air the WHO [World Health Organization] recommendations for gonorrhea. And gonorrhea, okay, you could say that it is quite trivial, but there are millions of people per year who get it. And in the meantime gonorrhea has become resistant to quite a few other successive treatments. . . . There is a new epidemic coming up of resistant gonorrhea." The rise in drug-resistant gonorrhea is causing concern in many parts of the world already, including the United Kingdom. While many of those who contract the infection experience no symptoms, left untreated it can lead to infertility in women, and it increases the risk of HIV transmission. In April 2016, Public Health England (PHE) reported a rise in the number of cases of gonorrhea that were highly resistant to one of the two drugs used to treat it, azithromycin. Gwenda Hughes, head of the Sexually Transmitted Infections Section of PHE, said at the time that "if strains of gonorrhea emerge that are resistant to both azithromycin and ceftriaxone [the second treatment option], treatment options would be limited, as there is currently no new antibiotic available to treat the infection." In July 2017, the WHO issued a warning that strains of gonorrhea that were difficult or impossible to treat had now been found in seventy-seven countries, and that this figure could be just the tip of the iceberg of a much more pervasive global spread of the disease.

Piot also spoke about the resistance problems he has seen with malaria, which is caused by a parasite. "We have seen, coming out of Southeast Asia, malaria that is no longer treatable with chloroquine, which was a very cheap effective treatment. Now we have [the drug] artemisinin, [derived from] Chinese traditional medicine. And now we have resistance developing there as well. If that [resistance] moves to Africa, we're in deep, deep trouble. That could really mean millions of deaths." He thinks it is mostly a matter of time until artemisinin-resistant malaria appears in Africa.[1] "I'm very concerned that that's not considered a Public Health Emergency of International Concern. That's a specific legal term under the international health regulations. That's what was used for Ebola, now for Zika. And I think this is one as well. Because it has a number of legal implications—this classification triggers a duty to assist and report; some of the sovereignty of countries is minimized, as the health emergency is considered of international concern, and addressing it is viewed as a global public good. If my neighbor's house is on fire, I have the right to go into my neighbor's house and put out the fire."

With his team, Neil Woodford, head of the antimicrobial resistance unit at Public Health England, analyzes samples to spot trends in emerging resistance. Woodford spoke to us about his experience working directly with superbugs. "The biology of bacteria never ceases to amaze the scientists. There are many examples of resistance where previous generations of scientists have proclaimed that resistance will be impossible—or so difficult that it will not emerge in the clinic. And we have then gone on to see rising rates of resistance to those very same antibiotics. . . . If history teaches us anything, it teaches us never to say never. The most recent example would be the transferable colistin resistance which was reported for the first time in November 2015 in China, and has subsequently been found around the world as people have started looking for it." The development of a transferable resistance gene means that the bacteria are able to quickly and easily transfer between each other the traits that enable them to beat colistin. Woodford thinks that "this has the potential to be very worrying. We've seen over the last decade or more how bacterial strains with resistance to carbapenem antibiotics have emerged, diversified, and spread to become what's now seen as the pinnacle of our resistance problems. . . . People are [now] finding bacteria that are resistant to the carbapenems and that also have this transferable colistin resistance. Not in large numbers yet, but maybe that's the direction of travel."

Antimicrobial Resistance and International Awareness

Antibiotic resistance has been recognized as a problem ever since antibiotics were first discovered. Fleming himself said, in an interview in 1945 not long after winning the Nobel Prize, "The thoughtless person playing with penicillin treatment is morally responsible for the death of the man who succumbs to infection with the penicillin-resistant organism." This was an early warning that unnecessary or inappropriate use could lead to these incredible treatments becoming ineffective. However, it is only in recent years that antimicrobial resistance has gained visibility on the international political agenda.

The momentum that has developed is the result of the collective work of a huge number of individuals and organizations, some of whom we have highlighted in previous chapters. An important step came in 2009, when the Swedish government made the crisis of antimicrobial resistance an area of focus as part of its presidency of the European Union. Scandinavian countries have led the world in recognizing the problem of drug-resistant infections and calling for global action to address it. The momentum began to gather strength after the WHO World Health Assembly ratified a Global Action Plan on drug-resistant infections in 2015, and additional commitments for national and global action were made at the 2015 meeting of the G7 (a group of countries from the world's seven largest economies).

It was also becoming clear that this problem required attention from areas outside the health sphere—particularly from economic and finance ministries. Meetings of the 2016 G20 (a group of countries comprising twenty of the world's largest economies) and then the United Nations (UN) brought the issue into focus. Then UK prime minister David Cameron commissioned the Review on Antimicrobial Resistance and championed the fight against drug-resistant infections. Cameron was the first head of government of a G20 nation to speak publicly and forcefully about the topic. When we talked with him in October 2016, he admitted that he knew little about the subject before he came to office. It was brought to his attention by his most senior medical adviser: "Genuinely it was the chief medical officer, Dame Sally Davies, who I had a regular meeting with. . . . She said, 'Look, I just want to bend your ear about this particular problem,' something which I had almost no knowledge of at all. If someone had said 'What is antimicrobial resistance?' I really would not have known."

This honest reaction shows how widespread the lack of awareness of antimicrobial resistance is—whether it is a top politician or everyday members of the public. Most people do, however, quickly grasp the severity of the situation when they learn about it. Cameron certainly did: "I suddenly saw this is an enormous problem but also an enormous opportunity for Britain to play a leading role, because we have this great position of being on the G7 and the G20, and through the other organizations that we're part of, we can be agents for global change. And it seemed to me that this is the perfect example of something that needed international action."

Cameron went on to explain that once Davies had brought home to him the gravity of the problem, he was able to get it on the global agenda with the help of a group of senior government officials known as "sherpas."[2] One such official was Tom Scholar, then the prime minister's adviser on European and global issues. Scholar, who was particularly involved in the discussions on antimicrobial resistance at the G20 in 2016, believes that the focus on economics helped put the issue on the agenda of the heads of government, the so-called "sherpas' track": "The sherpa world is political, and it's all about which leader wants what initiative. . . . But putting it through the finance track for a year turned it into a technical, analytical, economics issue; . . . it gave it credibility in the sherpa/heads of government world, which then helped support the political case which the UK government was making for it."

As important as it was for the UK government to invest political capital in this topic at the highest levels, action from the United Kingdom alone would not have led to an agreement. This was particularly true at the UN, where the high-level meeting in 2016 led to a political declaration on antimicrobial resistance (as discussed in Chapter 7). Delicate diplomacy was required to unite the common interests of 193 member countries and achieve a consensus. This was only the fourth time in the history of the UN that a health issue had been discussed at this level.

The Mexican ambassador to the UN, Juan José Gómez Camacho, led the daunting process of negotiating the declaration. When we spoke with him in December 2016, he described how important the work on the economics of drug-resistant infections was for making the case for action: "The first thing that I was looking into and the first thing that I was attracted to was precisely the economics. When Jim [O'Neill] produced his report saying basically that by 2050 the economic cost of this is going to be $100 trillion, within 2.5 and 3.5 percent of global GDP, then you can get it." But despite the enormous scale of the threat, Gómez Camacho needed to navigate the often vexed relationships between groups of countries within the UN, while overcoming the lack of understanding of the subject within the diplomatic community. Although UN diplomats must often negotiate deals on topics they know relatively little about, the more obscure and technical the subject, the more tempting it is for them to fall back into ideologically entrenched stances. Before the formal discussions and negotiations began, Gómez Camacho convened meetings of small numbers of permanent representatives (the top diplomats at the UN) from about forty countries in different regions, where he explained the issues. He considered it essential to educate delegates about the global importance of antimicrobial resistance, to think beyond the typical North-South ideological positions. During critical points in later negotiations, these earlier discussions helped guide the teams: "They understood that what we were doing was incredibly meaningful because it was incredibly threatening and dangerous to all," explained Gómez Camacho.

After having led the 2016 high-level meeting to its successful conclusion, with a detailed declaration covering most aspects of the problem of antimicrobial resistance, Gómez Camacho has remained committed to advancing this agenda at the UN. In the summer of 2017, along with his counterparts from the United Kingdom, China, Ghana, and South Africa, he formed an influential group of very senior diplomats who have pledged to support the UN's work on antimicrobial resistance in the run-up to the 2018 meeting, when it will once again feature on the agenda of the UN General Assembly. This return to the General Assembly will also be guided by the work of an Interagency Coordination Group (IACG) that was established by the UN secretary general in early 2017 and draws input from more than twenty international agencies and civil society groups.

This high-level political engagement at the UN has been mirrored by growing attention from the world's political leaders in other major forums, such as the World Health Assembly, G7, and G20.

At the 2016 G7 meeting, which was chaired by Japan, world leaders recognized the market problems for new drug development and called on the international community to rectify the issue. They also recognized the need to increase access to antibiotics as well as improving stewardship in both humans and animals. Balancing access and excess is essential but challenging.

The 2016 G20 meeting, chaired by China for the first time, also delivered a very promising statement on drug-resistant infections, which recognized that antimicrobial resistance "poses a serious threat to public health, growth and global economic stability," committed to "developing evidence-based ways to prevent and mitigate resistance, and unlock research and development into new and existing antimicrobials," and called on the WHO, FAO (Food and Agriculture Organization), OIE (World Organisation for Animal Health), and OECD (Organisation for Economic Co-operation and Development) to report back in 2017. Attention from the larger group of countries belonging to the G20 is vital to finding a global solution.

More progress occurred in 2017, with the G20 (its rotating presidency having transferred to Germany) reiterating calls for development of new antibiotics, as well as increased access to vaccines and diagnostics. The communiqué from the leaders' summit in Hamburg in July 2017 contained some of the strongest political statements on the issue to date, including a commitment to reduce antibiotic use in agriculture. We will see in the upcoming years if the intentions set out in that communiqué lead to significant reductions in the use of antibiotics in livestock in the G20 countries. A more concrete development at the G20 meeting occurred in the area of research and development. There was a call for an international research and development (R&D) collaboration hub that would work to "maximize the impact of existing and new anti-microbial basic and clinical research initiatives as well as product development." This initiative could help better monitor and coordinate the investment already being made by some governments and organizations, such as the Combating Antibiotic-Resistant Bacteria Biopharmaceutical Accelerator (CARB-X), discussed in Chapter 4, and the Global Antibiotic Research and Development Partnership launched in 2016 as a collaboration between the WHO and the Drugs for Neglected Diseases Initiative (a nonprofit with headquarters in Switzerland). The Berlin-based R&D hub has the potential to encourage more countries to become involved in research and development relating to antimicrobial resistance, as well as advancing discussions about how early stage incentives for R&D can be supported by new rewards for the development of antibiotics and alternative treatments. Having a wide range of active participants ensures that research efforts are sufficiently broad and reflect different areas of global need.

This overview indicates how much international awareness of antimicrobial resistance has increased over the past few years, with commitments to action from governmental officials. Although the political progress is positive, however, difficulties are likely to surface when countries actually attempt to make significant investments and change behaviors. Countries must work to live up to the promises made in these agreements, and there are signs that this is beginning to happen. We hope that this book will stimulate more attention to the specific proposals we have outlined.

Political Recommendations

Having generated political momentum, it is critical to make tangible progress toward solutions. This stage is difficult, as people tend to pay attention to the most dire global crises rather than those that are still developing. With so many urgent topics to consider, drug-resistant infections are in danger of not being seriously addressed until the problem becomes even worse. Tedros Adhanom Ghebreyesus, current director general of the WHO, has confirmed the organization's commitment to tackling rising drug resistance, but he has also underlined the challenges of doing so when so many health-care systems around the world remain weak and underfunded.

The tendency to be reactive could be countered if the right organizations held countries accountable for progress. The UN's 2016 declaration laid the groundwork by establishing an Interagency Coordination Group on Antimicrobial Resistance in March 2017, co-chaired by the UN and WHO, to report on progress in individual countries. This group must be given sufficient authority to pressure countries to deliver on promises made in the UN declaration. A UN special envoy on antimicrobial resistance could also raise awareness and provide external pressure. Additionally, an eminent figure from the arts or sports world could take a leading role in raising awareness among the general public. Two recent examples are Leonardo

DiCaprio's appointment in 2014 as a UN Messenger of Peace focusing on climate change, and Angelina Jolie's role as a special envoy on major crises that result in mass population displacements. A high-profile figure explaining the problem of resistance could have a huge impact in raising awareness and maintaining political pressure on policymakers.

Once international policy decisions are made, an operational plan will have to be put in place. We believe that it might be necessary to form a new entity to answer some key questions about how to proceed. In February 2017 the WHO published a list of priority pathogens for which antibiotics are urgently needed. A new entity could use this list as a guide in deciding how to give out rewards for new drugs, diagnostics, and vaccines. The G20 R&D hub described above might play this role. We recommend that the entity also look at the question of antibiotic use in farm animals and fish. The World Economic Forum set up a working group in early 2017 to study operational issues, in particular how rewards would be given out and managed, and it is due to report back at the 2018 meeting.

A number of specific policy objectives also require further international action. The initial work by the G20 recognized problems in the drug development market, but subsequent G20 meetings need to determine what interventions could create incentives for developing new drugs and diagnostics, and how to fund these interventions. This forum is vital, according to David Cameron, who told us that the G20 "is capable of actually taking a decision and taking action." While international agreements progress, individual countries should set up pilot systems to test how incentives such as our proposed market entry rewards system (described in Chapter 4) could work in practice. Such pilot programs would help to inform coordinated global action to deliver the right incentives to create new breakthrough antibiotics.

Although there are challenges on both the supply and the demand side, we believe that more work is required on the demand side at the international level. This is especially true in the area of rapid diagnostics, where little attention has been given to economic considerations. International discussions must take place on how to correct the market problems to encourage their development and use. The European Union's Innovative Medicines Initiative took the first steps in this direction, beginning a consultation in the summer of 2017 about how best to encourage new diagnostics that have the potential to curb antimicrobial resistance. A more ambitious approach would be for a group of high-income countries to mandate the use of diagnostics, as a way to increase investment in this area.

The overuse of antibiotics in agriculture and aquaculture also needs more attention. The WHO included the issue as part of its Global Action Plan, with member countries required to develop National Action Plans to tackle drug-resistant infections and consider animal usage. This work has been supported by the FAO and the OIE. More work is needed on international agreements as well as national programs to reduce use of antibiotics in farming, taking into account differences between countries in farming systems and climate. As discussed in Chapter 7, targets for antibiotic use could be determined by national governments, based on international agreements that would encourage countries to set ambitious targets and would monitor how much progress each country was making. This system of targets would ideally be agreed to by all 193 countries in the UN, but if that were not possible, then at least the G20 countries should participate, since they produce about 80 percent of the world's meat. Giving individual countries some autonomy in choosing target levels and mechanisms to achieve them—for example through taxes, regulation, or subsidies—would help drive positive changes.

Investing in Solutions

After we craft international agreements and decide on what interventions to undertake, we face a daunting question: Who will pay for these interventions? Such funding should be considered an investment, not just a cost. In fact, it is an exceptionally good investment. The cost for all the interventions the Review on Antimicrobial Resistance recommended for tackling drug-resistant infections came to $40 billion over a decade (see Table 8.1), or about $4 billion a year.[3] The cost of inaction, a prospective cumulative impact of $100 trillion by 2050, puts this investment into perspective.

We have made suggestions throughout the book on how to fund these interventions. Money could be raised from a variety of sources, including national governments, international institutions, the pharmaceutical industry, taxes, and a voucher system.

Given the market failures discussed in Chapter 3, which have hindered the development of new drugs and other approaches, there is a strong rationale for government spending to cover at least part of the investment. If such an investment is not made

Table 8.1 Estimated global costs of tackling antimicrobial resistance over a period of 10 years (in US dollars)

Intervention	Cost	Time period
Promote the development of new antimicrobials and make better use of existing ones	$16 billion	10 years
Create a global innovation fund to support basic and noncommercial research in drugs, vaccines, and diagnostics	$2 billion	5 years
Roll out existing and new diagnostics and vaccines	$1–2 billion	per year
Conduct a global public awareness campaign	$40–100 million	per year
Total	**Up to $40 billion**	**per decade**

Data from Review on Antimicrobial Resistance.

now, governments will have to foot an even larger bill when drug-resistant infections become more of a global scourge. Reactive spending after a health crisis explodes is almost always more expensive than addressing it early on, as we have seen in the case of numerous other health threats, such as HIV, swine flu, and SARS. The example of Ebola, which we examined in Chapter 3, is a useful case study. Since the investment was not made in stopping the spread of Ebola before it reached a critical point, large amounts of money were needed to combat the problem later. The US government alone appropriated $5.4 billion in a single year to cover its internal and global response to Ebola. Wider economic impacts, particularly on the countries directly affected by the Ebola outbreak, included damage to economic growth, productivity, and tourism.

We estimate that drug-resistant infections kill around 1.5 million people every year across the world. If the numbers of people dying from such infections continues to rise, the economic impacts will also continue to grow, particularly for the tourism industries of those countries most affected. This is another reason why governments should be interested in doing something about this problem now. An intervention of $3–4 billion per year would be affordable for most governments, representing only around 0.05 percent of the G20's annual spending on health care (which totals around $7 trillion). The amount of money the United States set aside in a single year for Ebola is more than would be needed annually from the entire international community to start addressing the problem of antimicrobial resistance. The money is available, and governments need to make a commitment to using it to stem this growing crisis.

Funding from international institutions should also be earmarked for this purpose. We believe it is appropriate for the World Bank to take a leading role, because making headway on this health challenge would support long-term economic growth in countries that would not be able to make the investments themselves. The World Bank signaled its interest in this area with a comprehensive report on drug-resistant infections in September 2016, and it could follow through by providing additional funding and using its extensive international networks to help achieve global solutions. Other large charitable organizations, such as the Wellcome Trust and the Bill and Melinda Gates Foundation, could contribute in the same way as they have on HIV, Ebola, and other global health issues. These organizations not only have the ability to directly fund important interventions but also have the power and influence to find solutions and create the right environment for continued political engagement.

We argue that the pharmaceutical industry should also contribute to combating this problem. Many companies are not currently investing in the development of new antibiotics because they can make more money in other areas, despite the fact that the effectiveness of their other treatments depends on the availability of antibiotics. As we explained in Chapter 3, this has given rise to a free-rider problem, in which the majority of pharmaceutical companies are taking advantage of the minority of companies that are working to find new antibiotics for the future. Because the entire industry—not to mention patients and the public—will eventually suffer if nothing is done, we believe that the industry has an enlightened self-interest in taking a more long-term view by investing in new antibiotics, which would reduce the threat of rising drug resistance and help ensure the sustainability of their businesses.

One way to encourage companies to invest in antibiotics R&D is to introduce a small investment charge payable on total sales, with payment required for accessing markets, as we discussed in Chapter 4. Only those companies not investing in research on new antibiotics would pay the charge. This system would create incentives for some companies to invest in antibiotics development, while others would decide instead just to pay the small charge. The money raised would help to fund the market entry rewards or other incentives for companies that are investing in the necessary research. This approach provides a more stable funding stream, rather than reallocating existing funds from health budgets and development agencies. Politicians and priorities change. At the moment, drug-resistant infections are attracting more political attention than they ever have before, yet it is possible that some of that attention will dissipate as time goes by and other health crises arise. Therefore, consistent funding has a clear appeal. Although a global system would be best, this approach could also work within large individual markets or within a group of countries.

A tax on generic antibiotics, either those used for humans, for animals, or both, is another way of raising funds. Taxes not only bring in revenue but also influence behavior. To discourage smoking and its related health effects, for example, governments tax tobacco products. Money from taxes on antibiotics could either go into general revenue or could be used for specific programs. If a government wants to lower smoking rates, for example, it might decide to commit the tax money on tobacco to fund support services for those trying to quit smoking. A tax on antibiotics within individual countries or a group of countries could be used to raise funds for projects aimed at tackling drug-resistant infections. Such a tax could provide a more stable source of revenue than some other methods of raising funds.

While a tax on antibiotics might have positive behavioral effects by reducing the amount of antibiotics prescribed, it would actually be a very blunt tool. In high-income countries, such a tax would be unlikely to have a huge impact, since the price of antibiotics does not have much effect on prescription rates. In some low- and middle-income countries, however, antibiotics are more price sensitive. A tax might actually hinder necessary use of vital antibiotics if set too high, particularly if individuals pay for their own treatment. This situation is different from the case of a high tax on a product such as tobacco, since tobacco use is a lifestyle choice with no health benefits. If a tax on global generic antibiotic sales was used to fund all needed international interventions it would probably need to be about 9 percent, or 3.5 percent to only fund a system of market entry rewards. Achieving broad international consensus on such a large tax would be very challenging, and further analysis of the impact on affordability and access would be needed.

A tax on antibiotics used in animals might be a better solution, particularly in certain high-income countries. As discussed in Chapter 7, antibiotics are overused in agriculture and aquaculture to promote growth, as well as to mask poor husbandry practices. Given that most livestock and fish are reared for economic reasons, producers in a country instituting a tax on antibiotics would either reduce their levels of use or increase the price of their product. To further encourage better practices, the money from such a tax could be used to help fund infrastructure, support, and training that would enable farms to remain productive while decreasing antibiotic use.

A system of exchangeable vouchers, described in Chapter 4, could also be set up to provide a reward for new antibiotics. Although we do not recommend these vouchers as part of an ideal solution for funding initiatives to combat drug-resistant infections, we recognize that different systems may work better in different countries. One of the advantages of vouchers is that they do not require governments to directly raise the funds to pay for them—which might mean they could get more political traction in certain countries.

Many Groups Have Roles to Play

While international groups will continue to play a major role in reducing the impact of drug-resistant infections, national governments will always have key roles to play, both in encouraging further commitments at the international level, and in ensuring that each country delivers the changes needed. So far there have been many champions of this cause at the national level, as we discussed earlier in this chapter. This pressure needs to continue, and additional political capital will be required on the part of a number of governments. Senior leaders can only push for a limited number of items at international meetings. Antimicrobial resistance must continue to be at the top of the agenda for at least a few major countries, or this momentum could be lost. As David Cameron put it: "I think there's a technical side to this and a more emotional side. The technical side is that the way the G7 and the G20 work is that once you get something into those great big communiqués, there's a natural urge to update it at the next meeting. So it becomes [an item] on the agenda, which has a value in itself. What you really need, though, is the emotional side. These things will just become boxes to be ticked unless one, or two, or more countries decide that they want to inject political capital, and effort and momentum, and make it happen." We need leaders who are emotionally committed to championing this problem, understand the threat it poses, and are willing to invest their time and use their connections to make sure there is progress.

Governments need to go well beyond box-ticking. They must use their power to fix both the supply and demand problems and correct the market failures that have hindered work in this area. In addition to these major changes, there are relatively simple things that governments and health-care systems can do. One of these is to put a higher priority on infectious diseases and antimicrobial resistance by ensuring that doctors and scientists working in this area are paid competitive salaries. In the United States, for example, infectious disease doctors rank at the bottom in compensation across all medical fields. If governments and health-care systems recognize this problem as one of the greatest health threats humans face, then incentives to work in this area should not be so low. While attracting and keeping talented people is not purely a question of salaries, improving compensation could encourage more top doctors, researchers, and students to enter the field.

The private sector will also have a huge role in tackling this problem. While the pharmaceutical industry is vital in the fight against drug resistance, a number of other groups have key roles to play, including diagnostics developers, food producers, and farmers.

As mentioned in Chapter 7, the pharmaceutical industry released an action roadmap in September 2016 setting out the course to create new antibiotics, improve access to vital antimicrobials, reduce unnecessary use in humans and in agriculture, and clean up supply chains. Having set out these plans, the industry now needs to achieve them. Skeptics might point out that it is easy to make promises, but very hard to follow through on them. It is important that the industry regularly return to these commitments in an open and transparent way, in order to gauge progress. By continuing their involvement on the innovation side—whether by committing further direct resources to antibiotic development, or by working with governments on ideas such as an antibiotic investment charge—pharmaceutical companies would help to ensure not only that we get action, but that the industry is appropriately represented and a key part of the solution. Initiatives such as the Access to Medicines Foundation's forthcoming Antibiotics Benchmark, which will rank companies based on their contribution to the problem of antimicrobial resistance, could play an even greater role in helping to change companies' behavior in relation to the development and use of antibiotics.

Industry also needs to address the issue of pollution caused when active ingredients are released into local water systems during manufacturing, thus increasing the risk that bacteria will develop drug resistance. One short-term solution is to significantly increase the transparency of supply chains. If all companies published lists of where they purchase their active ingredients and carried out more robust checks of their suppliers, this problem could be quickly solved. In addition, a minimum standard for active ingredient disposal into the environment should be developed and implemented.

Developers of diagnostics need to continue to make the case for rapid diagnostics and to collaborate among themselves and with pharmaceutical companies. It is clear that to control drug-resistant infections over the long term, we will need to be able to identify them more precisely using such tools. The declaration published by diagnostics companies at the 2017 meeting of the World Economic Forum included commitments to build the long-term economic case for diagnostics as a public good, advocate for simplified regulatory processes and sustainable reimbursement, and improve access to diagnostics. This kind of collaboration needs to continue and to include the pharmaceutical sector, governments, and health-care providers.

On the agricultural side, food producers and farmers need to reduce their use of antibiotics. Farmers should work with veterinarians and the regulatory authorities to come up with new guidelines that would lower antibiotics levels while keeping animals healthy, and food producers and retailers should encourage and support such efforts. The food industry must also lead efforts to ensure that the antibiotics most critical for humans are not used in farm animals and fish. Regulators have a role here too, but the food industry needs to lead, supporting the long-term interests of everyone who relies on effective antibiotics, including its customers and employees.

Action among citizens, patients, and consumers is also vital to raise the level of awareness about drug-resistant infections and to encourage behavioral change to help slow their spread. The level of activism in this area is at the moment much lower than for other threats that affect similar, or even far smaller, numbers of people. This is partly because antimicrobial resistance is not a single disease, but also because there is no obvious "face" to the problem. Marc Mendelson, an infectious disease expert at the University of Cape Town, spoke with us in January 2017 about some of the lessons that could be learned from other health threats that had successfully mobilized action. "Human stories are very powerful. . . . There has to be a face to this, and that's been our problem with AMR [antimicrobial resistance] the whole way through: Firstly nobody understands the term. . . . Secondly even when you do understand that term it does not have an automatic socio-pictorial association, and that's a challenge." If the problem continues to get worse, such a "face" might develop naturally following a crisis, but we cannot wait for this to happen. Civil society groups should play a leading role here—and not just ones that are solely focused on drug-resistant infections. Other health-advocacy groups, including cancer organizations, whose members need working antibiotics to prevent infections when receiving treatment, should become involved. Explaining the problem to patients and members of the public in straightforward language, combined with launching campaigns that profile people who have suffered from resistant infections, could help to raise awareness and maintain pressure on governments, health-care providers, and industry to take action. Civil society groups could also raise money for basic research, following the lead of groups that champion the causes of cancer and Alzheimer's disease, to name but two. If there is success on the first point, and awareness of the scale of the problem does increase, the number of people looking to donate money to address the issue is likely to increase.

Academics are also crucial to solving this problem. The vast majority of the very early yet important science that underpins drug research is done by the academic community. This includes basic research, where the application of such knowledge is not immediately apparent. Much remains unknown about drug resistance, including how it develops, the likelihood of transmission to humans from environmental sources, and countless other issues. Without further research, we may never make the very significant breakthroughs required to counter this problem. The academic community, however, is not sufficiently involved, partly because work in this area is too often not seen as being new or exciting, and partly because it is harder to achieve citations—a marker of academic prestige. Increased funding from governments, health-care providers, and industry would help to sustain this important research. We also hope that the growing international attention on this topic will encourage more and more academics to recognize the gravity of this problem and resolve to do their part to fix it.

Another group of people who have a vital role in this fight are those that treat infections—doctors and veterinarians. These health professionals can be influential in improving how antibiotics are prescribed, and their training should provide a more thorough understanding of infections. In hospitals, more experts in infection prevention and control should be in senior roles, where they could emphasize its importance. Doctors in many parts of the world are under huge amounts of pressure from patients to prescribe antibiotics, as discussed in Chapter 6. Although some of them need to be better at saying no to patients if they think that an antibiotic is not needed, patients also bear some responsibility. As patients, we must learn not to expect or demand antibiotics for ailments that are unlikely to be caused by bacteria. These decisions can be difficult, since forgoing an antibiotic can in some cases result in a serious illness or death, whereas the benefits of not overusing antibiotics extend beyond the individual patient and are much harder to see. Use of rapid diagnostic devices would provide support for these decisions. Doctors should be involved in discussions about the development of these diagnostics, since they are likely to be the ones using the tests. However, with advances in diagnostic technology and increasing use of these devices, pharmacists and other healthcare professionals may increasingly take on the role of prescriber. Some pharmacies in the United Kingdom, for example, are already using rapid diagnostic tests to diagnose strep throat. As a result, it will become even more important for doctors and pharmacists to share expertise and findings.

Veterinarians will also have an important role in guiding farmers and food producers to reduce levels of antibiotic use, while ensuring that animals still receive treatment when needed. Vets from countries that have high levels of training on this topic and have made great strides forward could help other countries to catch up and optimize their use of antibiotics.

Finally, there are many things that all of us, including you, the reader, can do. One of the simplest yet most effective ways to prevent infections in the first place is to wash your hands, often and properly—something that very few people do. The CDC gives guidance on how to do this: first wet your hands with clean, running water, then lather your hands, including between your fingers, and clean under your nails. Next, scrub your hands for at least twenty seconds. A trick to remember how long you should scrub for is to hum the "Happy Birthday" song twice. Then rinse your hands and dry with a clean towel or air dryer. We should also not demand antibiotics when we are sick. We can buy meat and fish from sources that use antibiotics responsibly. And finally, we can spread the word. Civil society groups have a big role to play, but without the support of the general public—

through funding, membership, and consumer activism—these organizations will not have the power to ensure that this issue remains at the top of the agenda for years to come.

A Problem That Can Be Solved

Antimicrobial resistance is not an intractable problem. A lot of the solutions that are thought to be hard are already being done in other areas; a tax on the pharmaceutical industry to accelerate research incentives, for example, is no different from the taxes we levy on pollution or on the use of fossil fuels. Prescription practices can also be changed relatively easily—with a combination of creating incentives for new technology and ending the practice of physicians and veterinarians making more money the more they prescribe. One of the great things about this problem is that fixing it has much wider benefits: for public health, for the environment, and for sustainable farming and society in general.

Now what is needed is the political will to act and find the money to implement the incentives needed. The investment of $40 billion over ten years for the world to avoid a $100 trillion cost by 2050 should make any finance minister stand up and take note. The potential to prevent an increase from 1.5 million to 10 million deaths per year should make every one of us stand up and take note.

This problem cannot be solved with a single effort. The international community will have to monitor progress well into the future, as we do with other global threats, such as climate change. Bacteria will continue to evolve as we develop new ways to defeat them. However, with the right effort now, we can keep the superbugs from winning. We can ensure that antibiotics, one of the greatest medical discoveries the world has ever known, will continue to be effective long into the future.

Notes

1. When a Scratch Could Kill

This chapter was written by Jeremy D. Knox.

1. These statistics can be easily misunderstood; an average life expectancy of forty-seven does not mean that the "typical" person died during their forties. Many people have always lived into their seventies and eighties, but these historical averages are brought down by a volume of deaths earlier in life that is shocking by modern standards—especially among the very young and among women during childbirth.

2. Throughout this book, where we refer to *antibiotics*, we specifically mean the class of drugs used to treat bacterial infections. Where we use the broader term *antimicrobials*, we are referring to a wider group of drugs that includes antibiotics as well as antivirals (used to treat viral infections), antifungals (for fungal infections), and antiparasitics.

3. The program eventually yielded tyrothricin—a topical anti-infective that was first marketed in the United States in 1942 and has remained in use ever since—in combination with topical anesthetics, as the basis of over-the-counter throat lozenges.

2. The Rise of Resistance

1. Three times MIC is the level of antibiotic that clinicians aim to have in a patient's system when trying to combat an infection.

2. In regions where the rate of resistant infections was already above that level, the current rate was used.

3. Thankfully vancomycin resistance is not yet rising within MRSA infections, so the number of infections that are resistant to both methicillin and vancomycin is currently static. We do not know if this situation will last.

4. World Bank Estimates used 2007 US dollars, while the Review on Antimicrobial Resistance used 2013 US dollars. World Bank estimates were thus converted forward to 2013 values using USinflationcalculator.com in order to make the estimates comparable.

3. Failures in Tackling Drug-Resistant Infections

1. These figures do not include a new tuberculosis drug, which is more expensive, since it requires a six-month treatment time and so is not comparable to the others.
2. Generic drugs are copies of brand-name drugs that have exactly the same dosage, intended use, side effects, route of administration, risks, safety, and strength as the original drugs. These can only be sold when the patent and market exclusivity rights to the original drug have expired, and they are normally much cheaper than the cost of patented drugs.
3. Data are from the final report of the Review on Antimicrobial Resistance and FirstWord Pharma, https://www.firstwordpharma.com/.
4. In economics, these are called nonexcludable goods because people cannot be excluded from using them.
5. Based on the World Bank's estimate that $2.2 billion was wiped off the local economy, which is 3.7 million times greater than the annual salary in the region affected.

4. Incentives for New Drug Development

1. The Review on Antimicrobial Resistance's internal work to evaluate drugs in the 2015 pipeline was led by Neil Woodford, head of Public Health England's Antimicrobial Resistance and Healthcare Associated Infections Reference Unit and the Review's scientific adviser.
2. Patents normally last for a term of about twenty years, but because drugs cannot be sold until they have been approved by regulators, the de facto patent duration is shorter for pharmaceuticals than it is for other products.
3. Borrowing figures are from Bloomberg and are correct as of December 10, 2016.

5. Prevention Is Better than Cure

1. Based on a reading speed of 300 words per minute, as estimated in B. Nelson, Do You Read Fast Enough to Be Successful? Forbes / Entrepreneurs, 2012. Available at https://www.forbes.com/sites/brettnelson /2012/06/04/do-you-read-fast-enough-to-be-successful/#17c277e9462e.
2. See United Nations, Sustainable Development Goal 6, https://sustainable development.un.org/sdg6.
3. This figure is less than the 0.12 percent of combined GDP spent achieving the sanitation and hygiene portions of the Millennium Development Goals that were established in 2000, with a target of completion by 2015.

7. Agriculture and the Environment

1. In this chapter we often use the terms "agriculture" and "livestock" for simplicity, but the issues we discuss are applicable to all animals raised by humans for food, including cattle, sheep, pigs, poultry, and fish.
2. We are using the FDA definition of "medically important." This figure does not include fish.
3. Colistin was initially not widely used in humans because of its high toxicity and because other antibiotics were available when it came to market. However, as these other antibiotics have become less effective due to resistance, colistin has increasingly been used in humans as a last-resort antibiotic.
4. Antimicrobial use is stated in kilograms of active substance used. Industry growth is based on production volumes measured in metric tons round weight.

8. Next Steps

1. The first case of an artemisinin-resistant strain of malaria contracted in Africa was reported in February 2017 by Lu and colleagues.
2. Derived from the Sherpa ethnic group in Nepal, whose members are famous for guiding people through the Himalayas. Government sherpas are senior diplomats who guide leaders on international matters.
3. This does not include the costs of much broader health-care interventions that should receive investment anyway and would also have an impact on antimicrobial resistance, such as improving sanitation and health infrastructure.

Bibliography

For further information on the subject of this book, see the various reports issued by the Review on Antimicrobial Resistance, the commission established by then UK prime minister David Cameron in 2014 and chaired by Jim O'Neill. These reports and additional information about the Review are available at https://amr-review.org/. See in particular the following:

Tackling Drug Resistant Infections Globally: Final Report and Recommendations. 2016.

Infection Prevention, Control and Surveillance: Limiting the Development and Spread of Drug Resistance. 2016.

Vaccines and Alternative Approaches: Reducing Our Dependence on Antimicrobials. 2016.

Antimicrobials in Agriculture and the Environment: Reducing Unnecessary Use and Waste. 2015.

Safe, Secure and Controlled: Managing the Supply Chain of Antimicrobials. 2015.

Rapid Diagnostics: Stopping Unnecessary Use of Antibiotics. 2015.

Securing New Drugs for Future Generations: The Pipeline of Antibiotics. 2015.

Tackling a Global Health Crisis: Initial Steps. 2015.

Antimicrobial Resistance: Tackling a Crisis for the Health and Wealth of Nations. 2014.

Introduction

Chen, L., R. Todd, J. Kiehlbauch, et al. 2017. Notes from the Field: Pan-Resistant New Delhi Metallo-Beta-Lactamase-Producing *Klebsiella pneumoniae*—Washoe County, Nevada, 2016. *Morbidity and Mortality Weekly Report* 66(1): 33.

Chapter 1

Aminov, R. I. 2010. A Brief History of the Antibiotic Era: Lessons Learned and Challenges for the Future. *Frontiers in Microbiology* 1: 1–7.

Birchenhall, D. 2007. Economic Development and the Escape from High Mortality. *World Development* 35(4): 543–568.

Bucholz, K., and J. Collins. 2013. The Roots—A Short History of Industrial Microbiology and Biotechnology. *Applied Microbiology and Biotechnology* 97(9): 3747–3762.

Bud, R. 2007. *Penicillin: Triumph and Tragedy.* Oxford: Oxford University Press.

CDC on Infectious Diseases in the United States: 1900–99. 1999. *Population and Development Review* 25(3): 635–640.

Centers for Disease Control and Prevention. 1999. Healthier Mothers and Babies. *Morbidity and Mortality Weekly Report* 48(38): 849–858.

Chain, E., H. W. Florey, A. D. Gardner, et al. 1940. Penicillin as a Chemotherapeutic Agent. *Lancet* 236(6104): 226–228.

Cook, G. C. 2015. *Disease and Sanitation in Victorian Britain: Lessons for the "Third World."* Ely, UK: Melrose Books.

Cutler, D., and G. Miller. 2005. The Role of Public Health Improvements in Health Advances: The Twentieth-Century United States. *Demography* 42(1): 1–22.

Daniel, T. M. 2006. The History of Tuberculosis. *Respiratory Medicine* 100: 1862–1870.

Davies, S. C., E. Winpenny, S. Ball, et al. 2014. For Debate: A New Wave in Health Improvement. *Lancet* 384(9957): 1889–1895.

Dowell, S. F., B. A. Kupronis, E. R. Zell, and D. K. Shay. 2000. Mortality from Pneumonia in Children in the United States, 1939 through 1996. *New England Journal of Medicine* 342(19): 1399–1407.

Fairchild, A. L., and G. M. Oppenheimer. 1988. Public Health Nihilism vs. Pragmatism: History, Politics, and the Control of Tuberculosis. *American Journal of Public Health* 88(7): 1105–1117.

Fishman, J. A. 2007. Infection in Solid-Organ Transplant Recipients. *New England Journal of Medicine* 357(25): 2601–2614.

Fleming, Alexander. 1945. Penicillin. Nobel Lecture, December 11, 1945. Available at https://www.nobelprize.org/nobel_prizes/medicine/laureates/1945/fleming-lecture.pdf.

Global Observatory on Donation and Transplantation. 2014. Organ Donation and Transplantation Activities, 2014. Geneva: Organización Nacional de Trasplantes–World Health Organization. Download available at http://www.transplant-observatory.org/data-reports-2014/.

Institute for Health Metrics and Evaluation. 2010. Global Burden of Diseases, Injuries, and Risk Factors Study 2010, GBD Profile: United States. Available at https://www.healthdata.org/sites/default/files/files/country_profiles/GBD/ihme_gbd_country_report_united_states.pdf.

Kochanek, K. D., S. L. Murphy, J. Q. Xu, and B. Tejada-Vera. 2016. Deaths: Final Data for 2014. *National Vital Statistics Reports* 65(4). Available at https://www.cdc.gov/nchs/data/nvsr/nvsr65/nvsr65_04.pdf.

Levy, S. B. 1992. *The Antibiotic Paradox: How Miracle Drugs Are Destroying the Miracle.* New York: Plenum Press.

McKeown, T., and R. G. Record. 1962. Reasons for the Decline of Mortality in England and Wales during the Nineteenth Century. *Population Studies* 16(2): 94–122.

Medicine: 20th Century Seer. 1944. *Time,* May 15.

Neushul, P. 1993. Science, Government, and the Mass Production of Penicillin. *Journal of the History of Medicine and Allied Sciences* 48: 371–395.

Omran, A. R. 1971. The Epidemiologic Transition: A Theory of the Epidemiology of Population Change. *Milbank Memorial Fund Quarterly* 49(4): 509–538.

Public Health England. 2016. English Surveillance Programme for Antimicrobial Utilisation and Resistance (ESPAUR) Report 2016, November. Available at https://www.gov.uk/government/uploads/system/uploads/attachment_data/file/575626/ESPAUR_Report_2016.pdf.

Quinn, R. 2013. Rethinking Antibiotic Research and Development: World War II and the Penicillin Collaborative. *American Journal of Public Health* 103(3): 426–434.

Sneader, W. 2005. *Drug Discovery: A History.* Hoboken, NJ: Wiley.

Tomes, N. 2001. *The Gospel of Germs: Men, Women, and the Microbe in American Life.* Cambridge, MA: Harvard University Press.

White, R. J. 2012. The Early History of Antibiotic Discovery: Empiricism Rules. In: T. J. Dougherty and M. J. Pucci, eds., *Antibiotic Discovery and Development.* New York: Springer, 3–32.

Williams K. J. 2009. The Introduction of "Chemotherapy" Using Arsphenamine—The First Magic Bullet. *Journal of the Royal Society of Medicine* 102(8): 343–348.

Chapter 2

Abraham, E. P., and E. Chain. 1940. An Enzyme from Bacteria Able to Destroy Penicillin. *Nature* 146(3713): 837.

Aminov, R. I. 2010. A Brief History of the Antibiotic Era: Lessons Learned and Challenges for the Future. *Frontiers in Microbiology* 1: 134.

Baym, M., T. D. Lieberman, E. D. Kelsic, et al. 2016. Spatiotemporal Microbial Evolution on Antibiotic Landscapes. *Science* 353(6304): 1147–1151.

Centers for Disease Control and Prevention (CDC). 2013. Antibiotic Resistance Threats in the United States, 2013. Available at https://www.cdc.gov/drugresistance/pdf/ar-threats-2013-508.pdf.

Chambers, H. F. 2001. The Changing Epidemiology of *Staphylococcus aureus? Emerging Infectious Diseases* 7(2): 178–182.

Clissold, S. P., P. A. Todd, and D. M. Campoli-Richards. 1987. Imipenem/Cilastatin. A Review of Its Antibacterial Activity, Pharmacokinetic Properties and Therapeutic Efficacy. *Drugs* 33(3): 183–241.

Deloitte LLP. 2009. Review of the Highways Agency Value of Life Estimates for the Purposes of Project Appraisal: A Report to the NAO, April 15. London. Available at https://www.gov.uk/government/uploads/system/uploads/attachment_data/file/35209/review-value-life-estimates.pdf.

European Centre for Disease Prevention and Control (ECDC). Surveillance Atlas of Infectious Diseases. Available at https://atlas.ecdc.europa.eu/public/index.aspx.

European Centre for Disease Prevention and Control (ECDC) and European Medicines Agency (EMEA). 2009. The Bacterial Challenge: Time to React. Stockholm: ECDC. Available at https://ecdc.europa.eu/sites/portal/files/media/en/publications/Publications/0909_TER_The_Bacterial_Challenge_Time_to_React.pdf.

Filice, G. A., J. A. Nyman, and C. Lexau. 2010. Excess Costs and Utilization Associated with Methicillin Resistance for Patients with *Staphylococcus aureus* Infection. *Infection Control and Hospital Epidemiology* 31(4): 365–373.

Laxminarayan, R., A. Duse, C. Wattal, et al. 2013. Antibiotic Resistance—The Need for Global Solutions. *Lancet Infectious Diseases* 13(12): 1057–1098.

Lim, C., E. Takahashi, M. Hongsuwan, et al. 2016. Epidemiology and Burden of Multidrug-Resistant Bacterial Infection in a Developing Country. *eLife* 5: e18082; doi:10.7554/eLife.18082.

Mantle, S. 2015. Reducing HCAI—What the Commissioner Needs to Know. London: NHS England. Available at https://www.england.nhs.uk/wp-content/uploads/2015/04/09-amr-brim-reducing-hcai.pdf.

Rollo, I. M., and J. Williamson, with R. L. Plackett. 1952. Acquired Resistance to Penicillin and to Neoarsphenamine in *Spirochaeta recurrentis. British Journal of Pharmacology and Chemotherapy* 7: 33–41.

Thaden, J. T., S. S. Lewis, H. C. Hazen, et al. 2014. Rising Rates of Carbapenem-Resistant Enterobacteriaceae in Community Hospitals: A Mixed-Methods Review of Epidemiology and Microbiology Practices in a Network of Community Hospitals in the Southeastern United States. *Infection Control and Hospital Epidemiology* 35(8): 978–983.

Tremblay, M., and H. Vézina. 2000. New Estimates of Intergenerational Time Intervals for the Calculation of Age and Origins of Mutations. *American Journal of Human Genetics* 66(2): 651–658.

Ventola, C. L. 2015. The Antibiotic Resistance Crisis, Part 1: Causes and Threats. *Pharmacy and Therapeutics* 40(4): 277–283.

World Bank. 2017. Drug-Resistant Infections: A Threat to Our Economic Future. Washington, DC: World Bank. Available at http://documents .worldbank.org/curated/en/323311493396993758/pdf/114679-REVISED -v2-Drug-Resistant-Infections-Final-Report.pdf.

Chapter 3

Baize, S., D. Pannetier, L. Oestereich, et al. 2014. Emergence of Zaire Ebola Virus Disease in Guinea. *New England Journal of Medicine* 371: 1418–1425.

Centers for Disease Control and Prevention (CDC). 2016. 2014 Ebola Outbreak in West Africa—Case Counts. Available at https://www.cdc .gov/vhf/ebola/outbreaks/2014-west-africa/case-counts.html (updated April 13, 2016).

Chakma, J., G. H. Sun, J. D. Steinberg, et al. 2014. Asia's Ascent—Global Trends in Biomedical R&D Expenditures. *New England Journal of Medicine* 370(1): 3–6.

Davis, J. H., M. Landler, and C. Davenport. 2016. Obama on Climate Change: The Trends Are "Terrifying." *New York Times*, September 8.

Deak, D., K. Outterson, J. H. Powers, and A. S. Kesselheim. 2016. Progress in the Fight against Multidrug-Resistant Bacteria? A Review of U.S. Food and Drug Administration–Approved Antibiotics, 2010–2015. *Annals of Internal Medicine* 165(5): 363–372.

Foto, T., and C. Grady. 2009. How Much Is Life Worth: Cetuximab, Non–Small Cell Lung Cancer, and the $440 Billion Question. *Journal of the National Cancer Institute* 101(15): 1044–1048.

Hughes, J. M., M. E. Wilson, B. L. Pike, et al. 2010. The Origin and Prevention of Pandemics. *Clinical Infectious Diseases* 50(12): 1636–1640.

King, L. 2011. Neglected Zoonotic Diseases. In: Institute of Medicine (US) Forum on Microbial Threats, *The Causes and Impacts of Neglected Tropical and Zoonotic Diseases: Opportunities for Integrated Intervention Strategies.* Washington, DC: National Academies Press.

Kohanski, M. A., D. J. Dwyer, and J. J. Collins. 2010. How Antibiotics Kill Bacteria: From Targets to Networks. *Nature Reviews Microbiology* 8(6): 423–435.

Leach, M. 2015. The Ebola Crisis and Post-2015 Development. *Journal of International Development* 27(6): 816–834.

Luby, S. P., M. Agboatwalla, D. R. Feikin, et al. 2005. Effect of Handwashing on Child Health: A Randomised Controlled Trial. *Lancet* 366(9481): 225–233.

Moon, S., D. Sridhar, M. A. Pate, et al. 2015. Will Ebola Change the Game? Ten Essential Reforms before the Next Pandemic. The Report of the

Harvard-LSHTM Independent Panel on the Global Response to Ebola. *Lancet* 386(10009): 2204–2221.

Moser, P. 2005. How Do Patent Laws Influence Innovation? Evidence from Nineteenth-Century World's Fairs. *American Economic Review* 95(4): 1214–1236.

Niffenegger, J. P. 1997. Proper Handwashing Promotes Wellness in Child Care. *Journal of Pediatric Health Care* 11(1): 26–31.

Scannell, J. 2015. Four Reasons Drugs Are Expensive, of Which Two Are False. *Forbes*, October 13.

Zorzet, A. 2014. Overcoming Scientific and Structural Bottlenecks in Antibacterial Discovery and Development. *Upsala Journal of Medical Sciences* 119(2): 170–175.

Chapter 4

Brookings Institution. 2013. Incentives for Change: Addressing the Challenges in Antibacterial Drug Development. Meeting summary, February 27. Available at https://www.brookings.edu/events/incentives-for-change-addressing -the-challenges-in-antibacterial-drug-development/.

Frontier Economics. 2014. Rates of Return to Investment in Science and Innovation. Report to the Department for Business, Innovation and Skills, UK Government. Available at https://www.gov.uk/government/uploads /system/uploads/attachment_data/file/333006/bis-14-990-rates-of-return -to-investment-in-science-and-innovation-revised-final-report.pdf.

Habets, M. G. J. L., and M. A. Brockhurst. 2012. Therapeutic Antimicrobial Peptides May Compromise Natural Immunity. *Biology Letters* 8(3): 416–418.

Hoskin, D. W., and A. Ramamoorthy. 2008. Studies on Anticancer Activities of Antimicrobial Peptides. *Biochimica et Biophysica Acta (BBA)–Biomembranes* 1778(2): 357–375.

Liu, P., M. Müller, and H. Derendorf. 2002. Rational Dosing of Antibiotics: The Use of Plasma Concentrations versus Tissue Concentrations. *International Journal of Antimicrobial Agents* 19(4): 285–290.

McDonnell, A., J. H. Rex, H. Goossens, et al. 2016. Efficient Delivery of Investigational Antibacterial Agents via Sustainable Clinical Trial Networks. *Clinical Infectious Diseases* 63(4): S57–S59.

Morrison, C. 2015. Antibacterial Antibodies Gain Traction. *Nature Reviews Drug Discovery* 14(11): 737–738.

Outterson, K., and A. McDonnell. 2016. Funding Antibiotic Innovation with Vouchers: Recommendations on How to Strengthen a Flawed Incentive Policy. *Health Affairs* 35(5): 784–790.

Outterson, K., J. H. Powers, E. Seoane-Vazquez, et al. 2013. Approval and Withdrawal of New Antibiotics and Other Anti-infectives in the U.S., 1980–2009. *Journal of Law, Medicine, and Ethics* 41(3): 688–696.

Pastagia, M., R. Schuch, V. A. Fischetti, and D. B. Huang. 2013. Lysins: The Arrival of Pathogen-directed Anti-infectives. *Journal of Medical Microbiology* 62(10): 1506–1516.

Pew Charitable Trusts. 2015. Antibiotics Currently in Clinical Development. Available at http://www.pewtrusts.org/~/media/assets/2015/03 /antibioticsinnovationproject_datatable_march2015.pdf?la=en.

PhRMA. 2014. Medicines in Development 2014: Cancer, 2014 Report. Washington, DC: Pharmaceutical Research and Manufacturers of America. Available at http://phrma-docs.phrma.org/sites/default/files/pdf /2014-cancer-report.pdf.

Pinzone, M. R., B. Cacopardo, L. Abbo, and G. Nunnari. 2014. Duration of Antimicrobial Therapy in Community Acquired Pneumonia: Less Is More. *Scientific World Journal* 2014: 759138.

Sertkaya, A., A. Birkenbach, A. Berlind, et al. (Eastern Research Group). 2014. Examination of Clinical Trial Costs and Barriers for Drug Development. Report to the U.S. Department of Health and Human Services, Office of the Assistant Secretary for Planning and Evaluation, Washington, DC. Available at https://aspe.hhs.gov/system/files/pdf /77166/rpt_erg.pdf.

Sertkaya, A., J. Eyraud, A. Birkenbach, et al. (Eastern Research Group). 2014. Analytical Framework for Examining the Value of Antibacterial Products. Report to the U.S. Department of Health and Human Services, Office of the Assistant Secretary for Planning and Evaluation, Washington, DC. Available at https://aspe.hhs.gov/system/files/pdf /76891/rpt_antibacterials.pdf.

Wilcox, M. H., D. N. Gerding, I. R. Poxton, et al. 2017. Bezlotoxumab for Prevention of Recurrent *Clostridium difficile* Infection. *New England Journal of Medicine* 376(4): 305–317.

Chapter 5

Andremont, A., and T. R. Walsh. 2015. The Role of Sanitation in the Development and Spread of Antimicrobial Resistance. AMR Control 2015: 68–73. Available at http://resistancecontrol.info/wp-content/uploads /2017/07/10_Andremont-Walsh.pdf.

Armstrong, G. L., L. A. Conn, and R. W. Pinner. 1999. Trends in Infectious Disease Mortality in the United States during the 20th Century. *Journal of the American Medical Association* 281(1): 61–66.

Biran, A., W.-P. Schmidt, K. S. Varadharajan, et al. 2014. Effect of a Behaviour-Change Intervention on Handwashing with Soap in India (SuperAmma): A Cluster-Randomised Trial. *Lancet Global Health* 2(3): e145–e154.

Borchgrevink, C. P., J. M. Cha, and S. H. Kim. 2013. Hand Washing Practices in a College Town Environment. *Journal of Environmental Health* 75(8): 18–24.

Bowen, A. C., A. Mahe, R. J. Hay, et al. 2015. The Global Epidemiology of Impetigo: A Systematic Review of the Population Prevalence of Impetigo and Pyoderma. *PLoS One* 10(8): e0136789.

Boyce, J. M. 2001. Antiseptic Technology: Access, Affordability, and Acceptance. *Emerging Infectious Diseases* 7(2): 231–233.

Boyce, J. M., N. I. Havill, D. G. Dumigan, et al. 2009. Monitoring the Effectiveness of Hospital Cleaning Practices by Use of an Adenosine Triphosphate Bioluminescence Assay. *Infection Control and Hospital Epidemiology* 30(7): 678–684.

Carabin, H., T. W. Gyorkos, J. C. Soto, et al. 1999. Effectiveness of a Training Program in Reducing Infections in Toddlers Attending Day Care Centers. *Epidemiology* 10(3): 219–227.

Chawla, S. S., S. Gupta, F. M. Onchiri, et al. 2016. Water Availability at Hospitals in Low- and Middle-Income Countries: Implications for Improving Access to Safe Surgical Care. *Journal of Surgical Research* 205(1): 169–178.

Coffey, D., A. Gupta, P. Hathi, et al. 2014. Revealed Preference for Open Defecation: Evidence from a New Survey in Rural North India. *Economic and Political Weekly*, September 20.

Cronk, R., and J. Bartram. 2015. Water, Sanitation and Hygiene in Health Care Facilities: Status in Low- and Middle-Income Countries and Way Forward. Geneva: World Health Organization. Available at http://apps .who.int/iris/bitstream/10665/154588/1/9789241508476_eng.pdf?ua=1.

Curtis, V., R. Aunger, and T. Rabie. 2014. Evidence That Disgust Evolved to Protect from Risk of Disease. *Proceedings of the Royal Society B: Biological Sciences* 271(Suppl. 4): 131–133.

Dancer, S. J. 2009. The Role of Environmental Cleaning in the Control of Hospital-Acquired Infection. *Journal of Hospital Infection* 73(4): 378–385.

Doebbeling, B. N., G. L. Stanley, C. T. Sheetz, et al. 1993. Comparative Efficacy of Alternative Handwashing Agents in Reducing Nosocomial Infections in Intensive Care Units. *American Journal of Infection Control* 21(1): 48–49.

Hancock, R. E. W., A. Nijnik, and D. J. Philpott. 2012. Modulating Immunity as a Therapy for Bacterial Infections. *Nature Reviews Microbiology* 10(4): 243–254.

Hutton, G. 2012. Global Costs and Benefits of Drinking-Water Supply and Sanitation Interventions to Reach the MDG Target and Universal Coverage. Geneva: World Health Organization. Available at http://www.who .int/water_sanitation_health/publications/2012/globalcosts.pdf.

Hutton, G. 2013. Global Costs and Benefits of Reaching Universal Coverage of Sanitation and Drinking Water Supply. *Journal of Water and Health* 11(1): 1–12.

Hutton, G., and M. Varughese. 2016. The Costs of Meeting the 2030 Sustainable Development Goals Targets on Drinking Water, Sanitation, and Hygiene. Technical Paper, Water and Sanitation Program, World Bank. Available at http://documents.worldbank.org/curated/en /415441467988938343/pdf/103171-PUB-Box394556B-PUBLIC-EPI -K8543-ADD-SERIES.pdf.

Judah, G., R. Aunger, W.-P. Schmidt, et al. 2009. Experimental Pretesting of Hand-Washing Interventions in a Natural Setting. *American Journal of Public Health* 99(Suppl. 2): S405–S411.

Laxminarayan, R., P. Matsoso, S. Pant, et al. 2016. Access to Effective Antimicrobials: A Worldwide Challenge. *Lancet* 387(10104): 168–175.

London School of Hygiene and Tropical Medicine. 2011. Contamination of UK Mobile Phones and Hands Revealed. Press Release, October 14. Available at https://www.lshtm.ac.uk/newsevents/news/2011/mobile phones.html.

Luby, S. P., M. Agboatwalla, D. R. Feikin, et al. 2005. Effect of Handwashing on Child Health: A Randomised Controlled Trial. *Lancet* 366(9481): 225–233.

Master, D., S. H. Hess Longe, and H. Dickson. 1997. Scheduled Hand Washing in an Elementary School Population. *Family Medicine* 29(5): 336–339.

McGowan, J. E. Jr. 2001. Economic Impact of Antimicrobial Resistance. *Emerging Infectious Diseases* 7(2): 286–292.

Niffenegger, J. P. 1997. Proper Handwashing Promotes Wellness in Child Care. *Journal of Pediatric Health Care* 11(1): 26–31.

Panigrahi, P., S. Parida, N. C. Nanda, et al. 2017. A Randomized Synbiotic Trial to Prevent Sepsis among Infants in Rural India. *Nature* 548: 407–412.

Pittet, D. 2001. Improving Adherence to Hand Hygiene Practice: A Multidisciplinary Approach. *Emerging Infectious Diseases* 7(2): 234–240.

Pofahl, W. E., C. E. Goettler, K. M. Ramsey, et al. 2009. Active Surveillance Screening of MRSA and Eradication of the Carrier State Decreases Surgical-Site Infections Caused by MRSA. *Journal of the American College of Surgeons* 208(5): 981–986.

Roberts, L., W. Smith, L. Jorm, et al. 2000. Effect of Infection Control Measures on the Frequency of Upper Respiratory Infection in Child Care: A Randomized, Controlled Trial. *Pediatrics* 105(4 Pt. 1): 738–742.

Rudan, I., C. Boschi-Pinto, Z. Biloglav, et al. 2008. Epidemiology and Etiology of Childhood Pneumonia. *Bulletin of the World Health Organization* 86(5): 408–416.

Ryan, M. A., R. S. Christian, and J. Wohlrabe. 2001. Handwashing and Respiratory Illness among Young Adults in Military Training. *American Journal of Preventive Medicine* 21(2): 79–83.

Snilstveit, B., and H. Waddington. 2009. Effectiveness and Sustainability of Water, Sanitation, and Hygiene Interventions in Combating Diarrhoea. *Journal of Development Effectiveness* 1(3): 295–335.

UNICEF/World Health Organization. 2015. Progress on Sanitation and Drinking Water: 2015 Update and MDG Assessment. Geneva: World Health Organization. Available at http://apps.who.int/iris/bitstream /10665/177752/1/9789241509145_eng.pdf?ua=1.

Waddington, H., B. Snilstveit, H. White, and L. Fewtrell. 2009. Water, Sanitation and Hygiene Interventions to Combat Childhood Diarrhoea in Developing Countries. Systematic Review no. 1, International Initiative for Impact Evaluation (3ie). Available at http://www.3ieimpact.org/media /filer_public/2012/05/07/17.pdf.

Weinstein, R. A. 2001. Controlling Antimicrobial Resistance in Hospitals: Infection Control and Use of Antibiotics. *Emerging Infectious Diseases* 7(2): 188–192.

Weinstein, R. A., D. R. Linkin, C. Sausman, et al. 2005. Applicability of Healthcare Failure Mode and Effects Analysis to Healthcare Epidemiology: Evaluation of the Sterilization and Use of Surgical Instruments. *Clinical Infectious Diseases* 41(7): 1014–1019.

Woerther, P.-L., C. Burdet, E. Chatchaty, and A. Andremont. 2013. Trends in Human Fecal Carriage of Extended-Spectrum β-lactamases in the Community: Toward the Globalization of CTX-M. *Clinical Microbiology Reviews* 26(4): 744–758.

World Health Organization. 2017. Immunization Coverage. WHO Fact Sheet, updated July. Available at http://www.who.int/mediacentre /factsheets/fs378/en/.

Chapter 6

Blaser, M. J. 2014. *Missing Microbes: How the Overuse of Antibiotics Is Fueling Our Modern Plagues*. New York: Henry Holt.

Borg, M. A. 2013. Prolonged Perioperative Surgical Prophylaxis within European Hospitals: An Exercise in Uncertainty Avoidance? *Journal of Antimicrobial Chemotherapy* 69(4): 1142–1144.

Center for Disease Dynamics, Economics and Policy. 2015. State of the World's Antibiotics, 2015. CDDEP: Washington, DC. Available at https:// cddep.org/wp-content/uploads/2017/06/swa_edits_9.16.pdf.

Devlin, H. 2016. Antibiotics Leave Children "More Likely to Contract Drug-Resistant Infections." *Guardian*, November 30. Available at https://www.theguardian.com/society/2016/nov/30/antibiotics-leave -children-more-likely-to-contract-drug-resistant-infections.

European Centre for Disease Prevention and Control. 2013. Point Prevalence Survey of Healthcare-Associated Infections and Antimicrobial Use in European Acute Care Hospitals, 2011–2012. ECDC Surveillance Report. Stockholm: ECDC. Available at https://ecdc.europa.eu/sites/portal /files/media/en/publications/Publications/healthcare-associated -infections-antimicrobial-use-PPS.pdf.

Gavi, the Vaccine Alliance. Advance Market Commitment for Pneumococcal Vaccines, Annual Report, 2015. Download available at http://www .gavi.org/funding/pneumococcal-amc/.

Government of India, Ministry of Health and Family Welfare. 2016. Health Minister Inaugurates 3-Day International Conference on Antimicrobial Resistance. Press Information Bureau, February 23. Available at http://pib .nic.in/newsite/PrintRelease.aspx?relid=136657.

Hallsworth, M., T. Chadborn, A. Sallis, et al. Provision of Social Norm Feedback to High Prescribers of Antibiotics in General Practice: A Pragmatic National Randomised Controlled Trial. *Lancet* 387(10029): 1743–1752.

Healthcare Improvement Scotland. 2008, updated 2014. Antibiotic Prophylaxis in Surgery. A National Clinical Guideline, SIGN 104, Scottish Intercollegiate Guidelines Network (SIGN). Available at http://www.sign .ac.uk/assets/sign104.pdf.

Leonard, R. 2014. In the Balance: GPs, Patient Care and Antibiotics. Blogpost, the Longitude Prize, August 19. Available at https://longitudeprize .org/blog-post/balance-gps-patient-care-and-antibiotics.

National Health Service, England. 2016. NHS England Chief Executive Simon Stevens Welcomes Eight Health Innovations Joining NHS Accelerator. Press release, November 14. Available at https://www.england .nhs.uk/2016/11/nia-innovations/.

National Health Service, England. 2016. NHS Innovation Accelerator: Sore Throat Test and Treat. Available at https://nhsaccelerator.com /innovation/sore-throat-test-treat/.

Pew Charitable Trusts. 2016. Antibiotic Use in Outpatient Settings. Available at http://www.pewtrusts.org/~/media/assets/2016/05/antibioticusei noutpatientsettings.pdf.

Roberts R. R., B. Hota, I. Ahmad, et al. 2009. Hospital and Societal Costs of Antimicrobial Resistant Infections in a Chicago Teaching Hospital: Implications for Antibiotic Stewardship. *Clinical Infectious Diseases* 49(8): 1175–1184.

Thaler, R. H., and C. Sunstein. 2009. *Nudge: Improving Decisions about Health, Wealth, and Happiness*, rev. exp. ed. New York: Penguin.

Wellcome Trust. 2015. Antibiotic Resistance Poorly Communicated and Widely Misunderstood by UK Public. Press release, July 29. Available at https://wellcome.ac.uk/press-release/antibiotic-resistance-poorly -communicated-and-widely-misunderstood-uk-public.

Chapter 7

Ågerstrand, M., C. Berg, B. Björlenius, et al. 2015. Improving Environmental Risk Assessment of Human Pharmaceuticals. *Environmental Science and Technology* 49(9): 5336–5345.

Declaration by the Pharmaceutical, Biotechnology and Diagnostics Industries on Combating Antimicrobial Resistance. 2016. Available at: https://amr-review.org/sites/default/files/Declaration_of_Support_for _Combating_AMR_Jan_2016.pdf.

Empire of the Pig. 2014. *Economist*, December 17. Available at http://www .economist.com/news/christmas-specials/21636507-chinas-insatiable -appetite-pork-symbol-countrys-rise-it-also.

Farrar, J. 2016. We Must Stop Squandering Our Precious Antibiotics. *Wellcome Trust News*, January 27. Available at https://wellcome.ac.uk/news/we -must-stop-squandering-our-precious-antibiotics.

Food and Agriculture Organization of the United Nations. 2016. At UN, Global Leaders Commit to Act on Antimicrobial Resistance. Press release, September 21. Available at http://www.fao.org/news/story/en/item /434147/icode/.

G20 Leaders. 2017. Shaping an Interconnected World, Combatting Antimicrobial Resistance, pp. 8–9. G20 Leaders' Declaration, July 7–8. Available at https://www.g20.org/Content/EN/_Anlagen/G20/G20-leaders -declaration.pdf?__blob=publicationFile&v=11.

Graham, J. P., J. J. Boland, and E. Silvergelb. 2007. Growth Promoting Antibiotics in Food Animal Production: An Economic Analysis. *Public Health Reports* 122(1): 79–87.

Guhal, S. 2014. If Leaders Fail, People Will Lead: Kofi Annan. *Times of India*, February 6. Available at http://timesofindia.indiatimes.com /interviews/If-leaders-fail-people-will-lead-Kofi-Annan/articleshow /29917567.cms.

Larsson, D. G. J. 2014. Pollution from Drug Manufacturing: Review and Perspectives. *Philosophical Transactions of the Royal Society B—Biological Sciences* 369(1656): 20130571.

Laxminarayan, R., T. Van Boeckel, and A. Teillant. 2015. The Economic Costs of Withdrawing Antimicrobial Growth Promoters from the Livestock Sector. OECD Food, Agriculture and Fisheries Papers, No. 78. Paris: OECD. Available at http://www.oecd-ilibrary.org/docserver /download/5js64kst5wvl-en.pdf?expires=1508960760&id=id&accname =guest&checksum=05D55B2EEC53D23E404DCE58AFA9EC83.

Leading Pharmaceutical Companies Present Industry Roadmap to Combat Antimicrobial Resistance. Press release, September 20. Available at http://www.ifpma.org/wp-content/uploads/2016/09/AMR-Roadmap -Press-Release_FINAL.pdf.

Levy, S. 2014. Reduced Antibiotic Use in Livestock: How Denmark Tackled Resistance. *Environmental Health Perspectives* 122(6): A160–A165.

Marshall, B. M., and S. B. Levy. 2011. Food Animals and Antimicrobials: Impacts on Human Health. *Clinical Microbiology Reviews* 24(4): 718–733.

National Resources Defense Council. 2016. Meat Raised without the Routine Use of Antibiotics Is Going Mainstream. NRDC Case Study 13-03-C, June. Available at https://www.nrdc.org/sites/default/files /antibiotic-free-meats-cs_0.pdf.

Norwegian Ministries. 2015. National Strategy against Antimicrobial Resistance, 2015–2020. Norwegian Ministry of Health and Care Services, publication number I-1164. Available at https://www.regjeringen.no /contentassets/5eaf66ac392143b3b2054aed90b85210/antibiotic-resistance -engelsk-lavopploslig-versjon-for-nett-10-09-15.pdf.

Ogle, M. 2013. Riots, Rage, and Resistance: A Brief History of How Antibiotics Arrived on the Farm. *Scientific American* guest blog, September 3. Available at https://blogs.scientificamerican.com/guest-blog/riots-rage -and-resistance-a-brief-history-of-how-antibiotics-arrived-on-the-farm/.

UK Government. 2016. UK on Track to Cut Antibiotic Use in Animals as Total Sales Drop 9%. Press release, November 17. Available at https://www .gov.uk/government/news/uk-on-track-to-cut-antibiotic-use-in-animals -as-total-sales-drop-9.

UK Parliament. n.d. Select Committee on Science and Technology, Seventh Report, Chapter 3: Prudent Use in Animals, para 3.7. Available at https://www.publications.parliament.uk/pa/ld199798/ldselect/ldsctech /081vii/st0706.htm.

United Nations. 2016. High Level Meeting on Antimicrobial Resistance. General Assembly, President of the 71st Session, press release, September 21. Available at http://www.un.org/pga/71/2016/09/21/press -release-hl-meeting-on-antimicrobial-resistance/.

World Bank. 2017. Drug-Resistant Infections: A Threat to Our Economic Future. Washington, DC: World Bank. Available at http://documents .worldbank.org/curated/en/323311493396993758/pdf/114679-REVISED -v2-Drug-Resistant-Infections-Final-Report.pdf.

Yi, L., J. Wang, Y. Gao, et al. 2017. *mcr-1*–Harboring *Salmonella enterica* Serovar Typhimurium Sequence Type 34 in Pigs, China. *Emerging Infectious Diseases* 23(2): 291–295.

Chapter 8

Doctors Fear Spread of "Super-gonorrhoea" across Britain. 2016. *Guardian*, 17 April. Available at https://www.theguardian.com/society/2016/apr/17 /gonorrhoea-will-spread-across-uk-doctors-fear.

European Commission. 2016. G20 Leaders' Communique Hangzhou Summit. G20 Leaders' Statement, September 5. Available at http://europa .eu/rapid/press-release_STATEMENT-16-2967_en.htm.

G20 Leaders. 2017. Shaping an Interconnected World, Combatting Antimicrobial Resistance, pp. 8–9. G20 Leaders' Declaration, July 7–8. Available at https://www.g20.org/Content/EN/_Anlagen/G20/G20-leaders -declaration.pdf?__blob=publicationFile&v=11.

Innovative Medicines Initiative, Diagnostics for Reducing Antimicrobial Resistance (AMR). Diagnostics Consultation Workshop, held June 19, 2017, Brussels. Links to agenda, presentations, etc., available at http://www .imi.europa.eu/news-events/events/diagnostics-consultation-workshop.

Lu, F., R. Culleton, M. Zhang, et al. 2017. Emergence of Indigenous Artemisinin-Resistant *Plasmodium falciparum* in Africa. Letter to the editor, *New England Journal of Medicine* 376: 991–993.

Penicillin's Finder Assays Its Future. 1945. *New York Times*, June 26, 21.

Safe Sex Reminder as Antibiotic Resistant Gonorrhoea Investigations Continue. 2016. Press release, Public Health England, April 17. Available at https://www.gov.uk/government/news/safe-sex-reminder-as-antibiotic -resistant-gonorrhoea-investigations-continue.

World Health Organization. 2017. Antibiotic-Resistant Gonorrhoea on the Rise, New Drugs Needed. WHO news release, 7 July. Available at http://www.who.int/mediacentre/news/releases/2017/Antibiotic-resis tant-gonorrhoea/en/.

World Health Organization. 2017. WHO Publishes List of Bacteria for Which New Antibiotics Are Urgently Needed. WHO news release, February 24. Available at http://www.who.int/mediacentre/news/releases/2017 /bacteria-antibiotics-needed/en/.

Acknowledgments

There are a large number of people we would like to thank for their help and support in researching, developing, and completing this book. First and foremost, we thank our former colleague Jeremy Knox for writing the first chapter, "When a Scratch Could Kill." Jeremy's knowledge of the history of antibiotic development is truly vast, and we are grateful to him for bringing his expertise to this chapter.

Jeremy and our other former colleagues Hala Audi, Olivia Macdonald, and Neil Woodford selflessly gave up a huge amount of time to discuss the book's ideas, read drafts, and ensure accuracy. Our knowledge of this issue, and many of the arguments in this book, originated with our work on the Review on Antimicrobial Resistance and the two years we spent working alongside these and other excellent colleagues. Their support throughout has been invaluable.

We would also like to thank Professor Dame Sally Davies for her continued help and wise guidance, and for kindly offering to write the book's foreword. Dr. John Rex was also kind enough to read a draft of the book and provide feedback.

So many people influenced our thinking on antimicrobial resistance over the past three years that it will not be possible to thank them all personally, but if you are one of the many people who shared your ideas on how to stop the rise of drug-resistant infections, then we thank you. We would like to extend thanks in particular to everyone we interviewed, whose ideas, often presented in direct quotes, helped us to clarify our thinking and helped bring this book to life: Manica Balasegaram, director of the Global Antibiotic Research and Development Partnership, Drugs for Neglected Diseases Initiative; Helen Boucher, director of the Infectious Diseases Fellowship Program at Tufts University School of Medicine; Derek Butler, chair of MRSA Action UK; David Cameron, former prime minister of the United Kingdom; Margaret Chan, former director general of the World Health Organization; Jeremy Coller, founder of the Jeremy Coller Foundation; Val Curtis, director of the Hygiene Centre at the London School of Hygiene and Tropical Medicine; Sally Davies, chief medical officer for England; Jeremy Farrar, director of the Wellcome Trust; Juan José Gómez Camacho, permanent representative of Mexico to the United Nations; Andrea Jenkyns, member of Parliament, United Kingdom; Nana Taoana Kuo, senior manager of Every Woman Every Child health team at the United Nations; Joe Larsen, director of the Division of CBRN Medical Countermeasures, Biomedical Advanced Research and Development Authority (BARDA), US Department of Health and Human Services; Joakim Larsson, Institute of Biomedicine, University of Gothenburg; Ramanan Laxminarayan, director of the Center for Disease Dynamics, Economics and Policy, Washington DC; Henry Lishi Li, London School of Hygiene and Tropical Medicine; Marc Lipsitch, professor of epidemiology, Harvard T. H. Chan School of Public Health; David McAdams, professor of business administration and economics, Duke University; Marc Mendelson, Division of Infectious Disease, Department of Medicine, University of Cape Town; Chantal Morel, research officer, London School of Economics and Political Science; Hồ Đặng Trung Nghĩa, head of Infectious Diseases Department, Pham Ngoc Thach University of Medicine; Jonathan O'Halloran, chief scientific officer, QuantuMDx; George Osborne, former chancellor of the exchequer, United Kingdom; Kevin Outterson, executive director of CARB-X;

Peter Piot, director of the London School of Hygiene and Tropical Medicine; Guy Poppy, chief scientific adviser to the Food Standards Agency, UK Government; John Rex, chief medical officer at F2G, Ltd.; Peter Sands, chair of the Commission on a Global Health Risk Framework, US National Academy of Medicine; Mark Schipp, chief veterinary officer of Australia; Tom Scholar, former adviser to the UK prime minister on European and global issues; Anthony D. So, founding director of the Innovation and Design Enabling Access Initiative, Johns Hopkins Bloomberg School of Public Health; Nicholas Stern, chair of the Grantham Research Institute on Climate Change and the Environment; Lucas Wiarda, director of marketing and head of the Sustainable Antibiotics Programme, DSM Sinochem Pharmaceuticals; Neil Woodford, head of Public Health England's Antimicrobial Resistance and Healthcare Associated Infections Reference Unit; and Bo Zheng, Institute of Clinical Pharmacology, Peking University First Hospital.

We would also like to thank Harvard University Press for asking us to write this book and all those who supported us there. In particular, we appreciate the continued guidance, patience, and excellent feedback from our editors, Janice Audet and Louise Robbins. Finally, we would like to thank our families and friends for supporting us as we undertook this challenging but exciting endeavor.

Index

Page numbers in *italics* indicate figures and tables.

things Pondered

things Pondered

A COLLECTION *of* POETRY AND VIGNETTES

BETH MOORE

FROM

THE HEART

of a

LESSER WOMAN

BROADMAN
&HOLMAN
PUBLISHERS

Nashville, Tennessee

Published by Broadman & Holman Publishers, Nashville, Tennessee

Ten-digit ISBN: 0-8054-2731-7
Thirteen-digit ISBN: 978-0-8054-2731-8

Previously published by Broadman & Holman under different packaging in 1997.

Dewey Decimal Classification: 242.64
Subject Heading: DEVOTIONAL LITERATURE\CHRISTIAN LIFE—POETRY

Unless otherwise noted, Scripture quotations are from the Holy Bible, New International
Version, copyright © 1973, 1978, 1984 by International Bible Society. Other version used is the
King James Version.

Library of Congress Cataloging-in-Publication Data

    Moore, Beth, 1957–
            Things pondered: from the heart of a lesser woman/Beth Moore
              p. cm.
            ISBN 0-8054-0166-0 (hardcover)
            ISBN 0-8054-2731-7 (repackage)
            1. Christian life—Baptist authors. 2. Moore, Beth, 1957–   .
        I. Title.
        BV4501.2.M5767 1997
        286'.1'092—dc21
        [B]   97–17520
                                                        CIP

3 4 5 6 7 8 9 10   10 09 08 07 06

IN MEMORY OF MY MOTHER

Whose feverish love for a word
fitly spoken endlessly inspires my own.
This offering is as much from her as to her.

"Mary treasured up all these things and
pondered them in her heart."

LUKE 2:19

Introduction

They were disappointed the moment they saw her. She even cried like a girl. They wanted a boy. That didn't mean they wouldn't love her. It just meant they were like every other set of Hebrew parents, hoping they would be the ones. They had no idea they had just given birth to the mother of God.

They named her Mary. The name meant "bitter." It was a name which may have described her calling, but it did not define her character. Little more than a child herself, she received the stunning news of the angel with grace and humility. Likely, Gabriel was relieved that he did not encounter the same insult he had with Zechariah. No wonder Elizabeth had exclaimed, "As soon as the sound of your greeting reached my ears, the baby in my womb leaped for joy. Blessed is she who believed . . . !" (Luke 1: 44–45). Her husband hadn't believed and not a single murmur from his lips had reached her ears in six months.

Mary had run to her the moment she had received the news. She hadn't hurried to her mother, father, friends, or fiancé. She had run to Elizabeth. How tender the God who shared with her through an angel that someone nearby could relate. The two women had one important predicament in common—questionable pregnancies, sure to stir up some talk. Elizabeth hadn't been out of the house in months. It makes you wonder why. As happy as she was, it must have

1

been strange not to blame her sagging figure and bumpy thighs on the baby. And to think she was forced to borrow maternity clothes from her friends' granddaughters. But maybe Elizabeth and Mary were too busy talking between themselves to pay much attention. Can you imagine their conversation over tea? One too old; the other too young. One married to an old priest; the other promised to a young carpenter. One heavy with child; the other with no physical evidence to fuel her faith. But God had graciously given them each other with a bond to braid their lives forever.

Women are like that, aren't they? We long to find someone who has been where we've been, who shares our fragile places, who sees our sunsets with the same shades of blue. Soul mates. They somehow validate the depth of our experiences. It is doubtful we have experienced much to which Mary could not relate. Through the course of her journey on this planet, she would experience fear, rejection, loneliness, disappointment, and heart break. She would know what it was like to have one child so entirely different from her others. She would battle sibling rivalry and yearn for her children to love one another. She would one day urge her oldest son to reach his potential, and although He would question her timing, He would nonetheless perform the miracle she requested. She would also confront her evolving role as the mother of an adult son. He would be no longer at her constant beck and call. And, ultimately, the woman highly favored by God would have to consider if the risk of loving was worth the risk of losing. Her small, suffering frame at the scene of her Son's death would prove to be testimony. One day, all those things would come. But in the meantime, as a young girl, she made something her practice that far surpassed her age or experience. She learned to catch a moment in her hand before it flew away and hold it tightly while she had the chance.

Luke's Gospel (2:8–18 KJV) tells it like this . . .

> And there were in the same country shepherds abiding in the field, keeping watch over their flock by night. And, lo, the angel of the Lord came upon them, and the glory of the Lord shone round about them: and they were sore afraid. And the angel said unto them, Fear not: for, behold, I bring you good tidings of great joy, which shall be to all people. For unto you is born this day in the city of David a Saviour, which is Christ the Lord. And this shall be a sign unto you; Ye shall find the babe wrapped in swaddling clothes, lying in a manger. And suddenly there was with the angel a multitude of the heavenly host praising God, and saying, Glory to God in the highest, and on earth peace, good will toward men. And it came to pass, as the angels were gone away from them into heaven, the shepherds said one to another, Let us now go even unto Bethlehem, and see this thing which is come to pass, which the Lord hath made known unto us. And they came with haste, and found Mary, and Joseph, and the babe lying in a manger. And when they had seen it, they made known abroad the saying which was told them concerning this child. And all they that heard it wondered at

those things which were told them by the shepherds.

And what about the young virgin mother?

But Mary treasured up all these things and pondered them in her heart. (v. 19)

Pondered. It's a wonderful word. It is the practice of casting many things together, combining them and considering them as one. In that moment a host of memories must have been dancing in her head. The angel's appearance. His words. Her flight to the hill country of Judea. Elizabeth's greeting. Their late-night conversations. The first time she noticed her tummy was rounding. Joseph's face when he saw her. The way she felt when he believed. The whispers of neighbors. The doubts of her parents. The first time she felt the baby move inside of her. The dread of the long trip. The reality of being full-term, bouncing on the back of a beast. The first pain. The fear of having no place to bear a child. The horror of the nursery. The way it looked.

The way it smelled. The way He looked. God so frail. So tiny. So perfect. Love so abounding. Grace so amazing. Wise men bowed down. Shepherds made haste. Each memory like treasures in a box. She gathered the jewels, held them to her breast, and engraved them on her heart forever.

The following pages are my responses to her worthy example. Words from a life absent of her lofty calling and excellent character. Experiences of an average woman, wife, and mother written to invite you to remember your own.

These are things pondered.

FAMILY TRACES

Wedding Bells

I should have been ready. It was an event I had prepared for all my life. But right at that moment, my wedding dress itched, my hair was bushier than my veil, and I couldn't get to a mirror for my bridesmaids. It was just as well. I was bound to be disappointed that the glamour of a film star was not included in the rental of my wedding gown.

This was not the way my sister, Gay, and I had played it. We had hosted at least a thousand rehearsal dinners with our Barbies and a bag of Fritos. Calendaring was certainly not the problem. I had always known that I would marry at Christmas of my twenty-first year. (That's how old Mother told me Barbie was. According to Mother, she had finished college before she married Ken.) I never had a wedding dress for my Barbie, but Mom had given me the most beautiful red velvet dress for her I had ever seen. I used it instead, which is exactly why she always had to have a Christmas wedding. Sure enough, it was December 30, and although I had on a traditional candlelight gown, my six bridesmaids were enchanting in their red Christmas dresses with capes, all carrying lanterns.

The groom also wasn't the problem. I knew he had to be from God. He came along the closest to the magical marrying age and my college graduation, and God forbid I would graduate without a proper betrothal. And more importantly, he was the spitting image of Ken. This qualification was not a conscious test at the time; however, it never occurred to me I'd marry anyone who was not tall, dark, and handsome. He certainly fit that bill, and he was the perfect companion for my Texas hair. Yes, the timing and the twosome seemed to be a match. Still, I had this sneaking suspicion welling up inside of me that I might be in for a slight shock.

The lightning bolt of a lifetime struck about one week later when we returned from our honeymoon to the old place we were renting from my father-in-law for free. Conspicuously absent was the portrait of me in my wedding gown over a blazing fireplace. In fact, the closest thing we had to a fireplace was the furnace to the left of the commode. The one I kept falling into because the seat was always up. There was no dishwasher, no garbage disposal, and no money. The only thing that house had plenty of was deer heads. They were everywhere. And they seemed to stare at me as if I had sold out. They did lead to the purchase of a secondhand dryer, however. Keith bought one the day after he came in with a friend and found our underwear dangling from their horns. I think that may have been the first time anyone had ever called me "sacrilegious."

My groom was not very impressed with playing grownups either. He took a cut in pay when he went from Daddy's allowance to hourly wages. He worked long hours out in the heat and worked with people who had no teeth. Then he'd walk in the door, say he was starving, and look at me as if he expected me to do something about it. Think as I may, I cannot for the life of me remember a kitchen in the Barbie Dream House. Then it hit me. I wasn't Barbie. He wasn't Ken.

This was no dream. And I wanted my mother. Now that I'm a parent, I have a feeling she wanted me, too, but she didn't let on. If I'd had a car that I didn't have to push to start, I might have been out of there. But there I was and there he was and even "making the best of it" seemed a dismal prospect. Had God not given us both the uncanny ability to laugh at inappropriate times, I don't think we could have made it. There were many months when we either laughed or cried, smooched or didn't speak. There was little in between.

We've suffered our share of bumps and bruises over the years since we drove off in our Barbie dream car and had a head on collision with reality. We've grown up a little and grown together a lot. We had entered marriage each carrying a deluxe, five piece set of emotional baggage, certain our own was heavier than the other's. We had expectations which exceeded the realm of possibility. I have finally forgiven Keith for not being Ken. I've almost forgiven myself for not being Barbie. And by the grace of God, we've made it in spite of ourselves.

Marriage is a serious matter. It is often embraced with less prayer than a college exam. I feel rather certain that the chief reason a Believer often enters marriage void of fervent prayer is because inherent in the asking is His right to answer us. And once we've made up our minds, a "no" or a "wait" from God is out of the question. Like us, you may have "accidentally" fallen into the marriage God had intended for you. Like us, you may have also suffered the penalty for not having built your relationship upon Him from the very beginning. And perhaps, like us, you're learning.

9

To marry without the blatant inclusion of Christ is to have entirely missed the point. In Ephesians 5:31 the Apostle Paul quotes the original wedding vows God spoke over the very first marriage all the way back in the Garden . . .

> For this reason a man will leave his father
> and mother and be united to his wife, and the two
> will become one flesh.

In Ephesians 5:32, thousands of years after God instituted marriage, He revealed its lofty purpose . . .

> This is a profound mystery—but I am talk-
> ing about Christ and the church.

Marriage is sacred. It was created to be the wedding portrait of Christ and His Bride hung over the blazing fireplace of judgment. A match made in Heaven, a contract signed in blood. In the bond of marriage, we are to stand at the altar of Sacrifice or we're not to stand at all.

> Colossians 1:16–17 gives us this assurance—

> By him all things were created . . . and in him
> all things hold together.

God alone created marriage. Adam slept through the entire ceremony. Eve came in late. It seems to me men are still sleeping through marriage, and women are still coming to their senses a little too late. God alone performed that ceremony, and He alone can hold it together.

Much of our disillusionment over marriage stems from the fact that we've been duped into believing that good equals easy. In other words, we often assume that if something is difficult, it can't be of God. Nothing has been more difficult for Christ than the marriage to His bride, yet Jude 24 says He'll present her to His Father with great joy! The Greek root word is *Agalliao*. It means "to show one's joy by leaping and skipping denoting excessive or ecstatic joy and delight!"[1] Just picture it. After all the ups and downs in the relationship, after all the marriage has cost Him, He'll act like a love-struck boy introducing his girl to his dad for the very first time. Why? Because He thinks she was worth it.

On the pleasant days of marriage, gaze across at your groom and conclude he is worth it. On the difficult days of marriage gaze up at your Groom and conclude He's worth it.

"A cord of three strands is not quickly broken."

ECCLESIASTES 4:12B

1. Spiros Zodhiates, ed. and compiler, *The Complete Word Study Dictionary of the New Testament* (Iowa Falls, Iowa: World Bible Publishers, 1992) 64.

Dear Bride to Be

Come to me, Dear Bride to be,
And kneel before My Throne
And I will share My heart with you
And make your house a home.
Listen well, lean closely
There are secrets at My feet—
The marriage you will soon begin
This Bridegroom will complete.

The man with whom you'll journey
Is your wedding gift from me
To teach you things beyond this world . . .
A precious mystery.
Bearing all these things in mind
You'll never lack for wealth
For through your union I will choose
To teach you of Myself.

Let him hold you tightly
And keep you safe from harm
Until I'll one day hold you
In My everlasting arms.
Let him wipe your tears away

And trust him with your pain
Until I wipe them all away
And Heaven is your gain.

Pray to love his tender touch
And want his gentle kiss
I grant you both my blessing
And ask you not to miss
The reason why I've chosen
For two halves to become one—
That you might see the Bride of Christ,
Sweet Daughter and Dear Son.

So make his home a refuge
He's to love you as I do
Until your mansion is complete . . .
A place prepared for you.
And if I should choose to leave you here
When I have called him home
Trust I'll be your husband near . . .
You'll never be alone.

Babies

Early one morning only eight weeks after the day we married, I heard an oddly controlled voice ascend from the kitchen. "It's positive, Beth. It came out positive."

That's impossible, I thought. After all, my doctor had assured me that birth control would be a waste of time. I was told that the fashion in which God had fearfully and wonderfully made me would make conception an impossibility without medical intervention. I was certain that the demands of marriage, i.e., cooking and cleaning, had made me understandably ill. I went to the kitchen to see for myself. There it was. The first sign of our first offspring—a jagged circle in the bottom of a glass test tube. I screamed. Then I laughed. Then I jumped up and down all over the linoleum. I had wanted a baby since the time I had been one. It was not until I threw my arms around my husband that I realized he was either terrified or had died standing up.

I set an extra place at the table that night and left it there until it was occupied. I immediately poked my stomach out as far as I could and practiced waddling in front of the mirror. And, most importantly, I enrolled in a child development course at the nearby community college. I planned to be the most wonderful mother in the world. Second only to the Blessed Mother herself. I never missed a single class and spoke as an authority on every point. I rolled my eyes at the ignorance of many of my classmates. It was clear they had not acquired the depth of experience I had while babysitting. Among the host of vows I made before my class, two were priority—I would never spank, and I would never say, "Because I said so." I was the fourth of five Army brats and the first one in line to ask "Why?" The tip of my father's index finger and those four words were stamped indelibly in my gray matter. I would be a far different kind of parent. I would simply explain things to my children. I would draw them up to my level and speak to them like little adults. I would patiently escort them to understanding and bask in the success of my modern methods. The course ended. I felt it was a shame the teacher could not give an "A+" on a report card. An "A" seemed so common.

I quickly enrolled in another class. It was called Lamaze Guide to Natural Childbirth. My husband had warmed up to the idea of starting a family due to the rapidly growing evidence that it was inevitable. He dutifully attended each one of our classes and fell asleep every night to my breathing exercises and stuck as closely to my side as my anti-stretch mark cream. I had highlighted my lime green Lamaze manual in fluorescent colors and worn the ink off the pages instructing how to experience minimal pain. I was not into pain. I was also far too contemporary to consider any drugs. I was expecting my first child at the exact moment the partum pendulum had swung from being knocked out to all natural. I was convinced that if I took one iota of pain reliever, my child would not only grow up to take drugs, she would likely sell heroin on the nearest corner. I hadn't made an "A" in child development for nothing. I made my husband promise that no matter how

I begged, he would not let me take any medication. I faithfully practiced my "hee hee" breathing and braced myself for the big day. And a big day it was sure to be. I had already caught a glimpse of myself in the stainless steel bathtub faucet while taking a bath. My navel looked like a helicopter pad. My stomach looked like the Astrodome, and my head and arms looked like they were down a few more exits. I took showers from then on.

One Sunday morning nine months and two weeks from our wedding, I awakened to my stomach doing abdominals independently. I whispered my suspicions into my husband's left ear, to which he responded by jumping straight out of the bed into his cowboy boots. I convinced him we had plenty of time to run by the church and let me teach a quick Sunday School lesson to my sixth graders on the way to the hospital. He conceded reluctantly, and we made a beeline for the church. I've never been known to teach a short lesson, and by the time we headed back home to grab my suitcase, the contractions were growing powerfully. Nothing in Houston, Texas, is vaguely close by, and our hospital was considered far even according to our standards. As we began our trek, I said to Keith, "It sure is a good thing I practiced my Lamaze as much as I did. This really hurts. Don't worry, though. I'm prepared and I feel in control." Minutes passed, and I was feeling a tad less controlled.

"Keith," I suggested, "you might try taking streets with no traffic lights. I don't think we ought to tarry at another one." By this time I was "hee hee" breathing.

A few minutes later, as my sweet husband began to panic, I said, "You stop again and you deliver this kid!" By this time my feet were on the dashboard. With eyes that could shoot torpedoes, I selected Keith's face as my focal point and announced emphatically, "I'M READY TO PUSH!" We pulled up in front of Rosewood Hospital on two wheels.

My husband flew in the door and said, "Forget the labor room. It's too late! Get her to delivery!" A worn-out nurse with a deadpan face fetched the doctor with the speed of molasses. I squeezed my knees together and gritted my teeth trying to hold it off until they got there.

The doctor arrived, examined me, and said, "Mrs. Moore, you are one and a half centimeters dilated." It was a long day.

I divorced Keith during every contraction and remarried him in between. I didn't know a soul could feel pain like that and live, and I was taking him with me if I didn't. My "hee hee" breathing had diminished to sobs of "he he made me do this!" Many hours and a large incision later, I gave birth. I don't know why they call it "expecting." There is nothing about it you could've expected. Her head was misshapen from the difficult delivery. Her head was bleeding from the forceps. Her cheeks were bruised from the pressure. And she was the most beautiful creature I had ever seen in all my life. Not many things in life are perfect, but every now and then, every once in a great while, there arrives a perfect moment. This was one of them.

That moment gave birth to the most joyful days I had ever experienced in my life. Keith and I thought all well-parented babies sat happily in their car seats, slept through the night, and bore signs of brilliancy from birth. Her name was Amanda. She talked nice and early and walked nice and late. As the infant turned into a delightfully creative toddler, her mind became a constant bed of fairy tales, and her nursery turned into a palace. An obvious "Save the Whaler" from her first words, she was born to crusade righteous causes, however off the wall, and stick up for the underdog. Deep spirited from the start, she once looked into the sky at three years old and with eyes spilling tears, said, "One lonely little cloud.I wish it had some friends." By senior high school she had changed wonderfully little.

I had never spanked her, nor had I ever said those four deplorable words. After all, I had made a vow. And I almost got away with it. That is, until I had that one fleeting thought when I glanced at her pink cheeks and pigtails as she swung at the neighborhood park, "I'm so good at this, it would be a shame for me not to do it again. After all, Keith wants another so badly." Just as the thought went through my head, the loudest clap of thunder I had ever heard nearly collapsed the sky. I know now that it was God having a good knee slapper.

I feel sure this was the moment when He summoned Gabriel and said, "Remember that little spirited spirit we've had around here all these centuries just waiting for the right mother? She just reported for duty."

I had no idea what God had planned. As usual, I had plans of my own. After all, planliness is next to godliness, is it not? Without informing my husband, I plotted my next pregnancy. I planned to be in the family way by Thanksgiving, just in time to give him the ultimate gift on Christmas Day. I couldn't imagine ever loving another little girl as much as I had my first, and I had always heard every Father needed a son, so I chose just the right formula of words and petitioned

God for a boy, in Jesus' Name . . . and in advance. I thought it would be simpler that way. My timing worked perfectly according to plan, and by Thanksgiving I was kissing Keith as he headed out the door by 7:00 a.m. and throwing up by 8:00 a.m. I wrapped a darling, but very manly pair of blue booties in crisp red and green wrap and stuck the special delivery all the way under the Christmas tree.

December 25 finally rolled around, and we gathered with extended family at my parent's home and began to exchange gifts. As forever his custom, my Father passed out every gift one by one. Finally, a single small package was under the tree. Having no idea what the present contained, he announced to my husband that it was addressed to him . . . from Santa Claus. Keith didn't waste much time tearing away the Christmas ribbons and paper, then stared totally puzzled at the fuzzy little booties. He finally looked at me across the room and mouthed the words, "Does this mean you want to?" I responded loudly, "No, my Darling, this means WE ARE!" He swept me up and swung me around while my family cried and I threw up. After the hysteria had died down, he inquired, "Just one more thing I'm curious about. Why are these blue booties?"

I quipped, "Because, Honey, I asked God for a boy, in Jesus' Name." How could he be so spiritually immature?

My tummy was swollen by the time I ate the second piece of pumpkin pie. I didn't care. I was filled with a complete satisfaction over well-executed plans. What could be better? A handsome husband, a precious daughter, a wonderful son, and a wiener dog named Coney Island! Boy, was I feeling sassy.

Seven months into the pregnancy the obstetrics nurse, who was a friend of mine, offered to do a sonogram for which she had been recently trained. I drank the four and a half gallons of water required and sloshed all the way to the doctor's office. She smeared the monitor with a clear, ice cold jelly then whacked it on my tummy pushing as hard as she could on my bladder. She positioned the television screen right in front of me, and we began to watch a performance of award-winning caliber. It was a miracle. The child flipped and turned and sucked its thumb. We saw everything! The eyes, the nose, the fingers and toes! Everything but one. After thirty minutes of Star Search, I finally said, "Have you been able to tell he's a boy yet?"

She responded reluctantly, "No, I just can't seem to get the right angle."

That baby didn't have an angle we hadn't seen. I retorted, "You're telling me a story. That baby's a girl!"

"You're absolutely right."

I went home, sat on the couch, and looked up at God. "You can change this, You know. You can either do it now, or You can do it right in front of the labor and delivery staff. I don't care when or how, just do it!" I added one little footnote. "But if You're going to do it, be sure You do it all the way." I didn't need anything else to worry about. I positioned myself very still to see if I'd feel anything. I didn't.

It wasn't that I didn't want another daughter. I love little girls! It's that I had shot off my big mouth to everyone this side of the Pecos, and I was far too pregnant to be sitting out on any limb. Somehow I had a feeling God had made up His mind. I broke the news apologetically to my husband. After a moment he chuckled and asked, "Am I supposed to be disappointed or something?" That was all I needed to hear. The next two months we ecstatically prepared for daughter number two. When I say we prepared, surely you're believing by now, we prepared! This time I passed on the Lamaze classes and entered the hospital with my "Say Yes to Drugs" T-shirt chanting those three magic words, "Just say yes!" The way I saw it, bugs and turnips were "natural" and I didn't like them either.

I never could tell that the anesthesia I had longed for had any effect at all until a few moments after I struggled and gave birth to that wonderful child. I awakened in the recovery room to my husband's tender words as he whispered in the tiny ear of his Christmas gift. She was dark complected like my "Ken" and had a white gauze cap on her head. He had her propped in his left hand and was holding up one little slat of the mini-blinds with the other. Her little eyes squinted as he promised, "It's a mighty big world out there, but don't you worry. It won't hurt you. Daddy will be right there." It was a perfect moment. My second.

That beautiful child was an angel from Heaven for two solid weeks when she abruptly opened her mouth and, to date, has not shut it. I began to have a vague recollection of that mighty clap of thunder on the pinnacle day of my piety. She sat way too early. She crawled way too early, and she ran way too early. She has yet to learn to walk. When her screams turned into vocabulary, she was finally able to put her frustrations into words. The first sentence out of her mouth is accurately recorded in her baby book—"Don't boss me!" Melissa has never been one to take sides. She came to take over.

We could always tell what kind of day it was going to be by the way her hair looked when she walked down the stairs in the morning. If she had experienced a rough night, it was going to be a rough day. We tried cutting her hair nice and short, but she still came down those stairs looking just like an angry banty rooster. Then came that fated day when she was four years old and demanded to go outside on a cold, rainy day. I said no. She asked why. I explained why. She said no. I explained again. She threw a fit again. "But WHY can't I, Mom?? You're just plain ole mean!" Just then, I felt a strange allergic reaction occurring around my mouth. It felt like ants were crawling all over my tongue. My lips began to blow. Then it happened. I couldn't control it. It just happened. In a compulsory refrain, I screamed, "BECAUSE I SAID SO!" Once was not enough. Every single one stored up inside me came out in a flash. I yelled it at the dog. I yelled it at the cat. I yelled it at the swing set. I yelled it sharp and flat. It was freedom! Liberation to my soul! Free at last! I fell exhausted on the couch, at which point my older child said with hands on her hips to her precocious little sister, "Because she said so. That's why." To which she responded, "Oh. OK." And she skipped away. When answers aren't enough, there is "Because I said so." Thanks, Major Dad.

Her stubbornness is half her charm. One day after Mother's Day Out, she announced to me proudly, "My teacher called me a different drummer!" I was furious. I knew what that teacher had really said. "Melissa sure marches to the beat of a different drummer." Just about the time I started to turn right back around and give that teacher a piece of my mind, I thought about the words Melissa had quoted. Her teacher said it one way. She heard it quite another. She didn't march to anyone's beat. She was the different drummer. And she still is.

I marvel at the ignorance of a mother who thought two daughters might be too much alike to enjoy. Their appearances were strikingly similar, but their personalities were gloriously unique from the start. One lived in a fairy tale. The other a tailspin. One cradled a red-and-black ladybug in her palm and said, "Oh, my pretty little ladybug." The other, half her size, looked over her shoulder and retorted, "It's dead, stupid." One aimed to please. The other aimed to push. One was my favorite. And so was the other.

Only God could have created a parent's love. Only to God can we all be His favorite, the "apple of His eye" (Ps. 17:8). How can we ever doubt that He loves us as much as another if we, as human parents, are capable of the same? What unfathomable depths of love must be in the heart of One who said,

> If you, then, though you are evil, know how to give good gifts to your children, how much more will your Father in heaven give good gifts to those who ask him!
>
> MATTHEW 7:11

I would give my children anything I could afford if only it wouldn't hurt them. He gave us everything He could afford—the riches of Heaven—His Son, and oh, how it hurt Him.

Today, with daughters I literally have to look up to, I arrive at the same conclusion I suspected many years ago when I danced with glee over my first babysitting job. I like children.

I Like Children

I like the way they're always full of surprises . . . how they have a mind of their own from the very beginning and arrive just in time to be two weeks late. I like the way they look like little strangers the moment you feast your eyes on them . . . totally unrecognizable yet freshly detached from your own body. I like the way they come to your hospital room in a plain white blanket, wrapped so tightly and with such precision you wonder if they'll have to wear it to college. I like how they look in their baby bed the very first time you tuck them in it—so small you decide they better sleep in your room.

I like the funny expressions they make while they're dozing and how they crack an awkward smile as if they've tagged an angel. I like the way they yawn with their whole bodies and how the stork bites on the backs of their necks are often as plain as day. I like the way they never go for the applesauce disguising the pureed liver. I like how they smell after their grannies bathe them and bring them to their mamas. I like the soft bristles of their brushes and how their hair looks when you first get it to part.

I like the way they love you more than anyone else on earth has ever loved you. I like how they quiet to your whisper after all your friends and relatives have desperately tried to calm them. I like the first time they reach their arms out to you. I like having the prerogative not to lay them down for a nap and rocking them instead for all three hours if you have a mind to.

I like the way they learn to entertain all the patrons at the restaurant with a spoon on the metal tray of a high chair. I like how they first say Mama and Dada with twenty syllables each. I like the dimples their knees make when they first learn to stand. I like how they learn to walk because they want to get to you. And, boy, do I like footie pajamas . . . until the next morning when no telling what is in the footie. I like the way they know they're going to Mother's Day Out the instant they wake up. And they're not in the mood. And I love sleepy hair. You know . . . how it looks all fuzzy on one side when they first wake up.

I like the sudden discovery of sentences as their thoughts take the form of endless, delightful vocabulary. I like how you nearly die laughing once you realize what they're trying to say. I like the way neighbors don't realize they've just been insulted because they can't understand a word out of their mouths. I like the way "R's" don't appear in their alphabet until they are at least five years old. I like their simple rules of socialization . . . move or I push . . . gimme or I bite.

I like how little girls think pink chiffon dresses are divine and little boys wear their cowboy boots with shorts. I like the way little girls prefer umbrellas and little boys—puddles. I like how they look on the first day of kindergarten—from the front. Not from the back. I like taking pictures of them with their friends every year on the first day of school . . . that is, until you come across that very first one in the drawer. And you cry. 'Cause it went too fast. And you can't go back. I like the way they know it's time to go even when Mommy doesn't agree. Because that's the way it should be.

I like how your children like you even better when they're grown. And how, if you're really lucky, they might have children of their own. And you can try it once more.

And maybe do a little better. Because I like children.

When my daughters were seven and ten and we were basking in the marvelous years between preschool and adolescence, I learned a life-changing lesson about prayer—God reserves the right to fill petitions you forgot to cancel long after you thought you changed your mind. All those years ago when I had asked God for a son, I assumed His answer was "no," not "wait." Boy, was I ever wrong. On February 14, 1990, my husband gave me a Valentine's gift that keeps on giving—a pint-sized, four-year-old orphaned boy. He was the most beautiful little guy I had ever seen in all my life. I've since arrived at the conclusion that God often makes children who are going to be extra work extra cute. At the time, however, his big brown eyes and inch-long eyelashes were simply selling points.

God had reserved room in our hearts and a room in our home for one more child. We were not looking to adopt a little boy. We were very satisfied with the size of our family. God had tendered our hearts over the plight of only one. His birth parents were married when he was born, but they soon gave up on each other and ultimately him. Sadly, the marriage of his second guardians also collapsed, and they sought a family to raise him. His name was Michael, but the girls soon nicknamed him Spud and it stuck. He was darling, very troubled, and the spitting image of his new Daddy.

These next words are not just phrases and rhymes. They comprise the events of a night that dramatically changed our lives. It was late that evening and our daughters were in bed. And, yes, it was a perfect moment. Our third. Our son.

The Adoption

I heard the front door open
My heart began to pound
I froze to see if it was them
Then I heard the sound

Of a tiny little four-year-old
Asking this strange man
"How come we gathered all my clothes?"
Did my husband have a plan

Exactly how to tell the child
On this awaited day
"You've left the only house you've known
And now you're here to stay"?

And what am I to say to him,
"Hello, my name is Mom"?
I was filled with insecurity
But my husband looked so calm

He diverted his attention
And didn't answer right away
He looked at me assuredly
"Let's just let him play."

After minutes crept and crawled away
He patted his right knee.

"Can I talk to you a moment, Child?
Would you sit right here with me?"

He stopped what he was doing
And crawled up on his lap
He looked straight into my husband's face
And dropped his baseball cap.

"Michael, do you have a Dad?"
My heart jumped in my throat.
A sadness swept that precious face
"No," he said, "I don't."

"Michael, I've been thinking
Since the first I saw of you
We've got a common problem
Is there something we can do?"

"I've everything that I could want
Upon my list but one.
It seems that you don't have a Dad
And I don't have a son."

"Whatcha say we strike a deal
And seal it with a shake?
I've thought it over carefully
Now, the choice is yours to make."

I remember well the boy's words,
Feet swinging as he sat—
After all this time without a Dad,
"It happens just like that?"

The man gave him a gentle nod
The boy's grin grew wet
As if he thought, "What's there to lose?"
He blurted out, "You bet!"

His hand appeared so fragile
In my husband's callused palm
Keith whispered, "There's a bonus here.
That lady's now your Mom!"

The handshake gave way to a hug
Tears came as no surprise
Transfixed, I watched my precious son
Be born before my eyes.

They ascended up our stairway
Suitcase tightly in his hand
My husband pointed to a door
And said, "Enter, little man."

Cautiously he took each step
Until he was inside
Bunk beds for boys, and lots of toys
Confused, his eyes grew wide

"Whose stuff is this?" The boy inquired
Not knowing what to do
"Go ahead and touch it, Child!
It all belongs to you."

The birthday of our special son
Should not seem strange or new

For if you have been born again
You've been adopted too.

For God so loves this aching world
He pulls us, good and bad,
Onto His lap and says to us,
"I want to be your Dad."

"You have no Heavenly Father
And you're not my son, it's true,
But there's room inside my family
A space made just for you."

"It's not a snap decision
It's the kind that takes the heart
But with a 'yes' all things are new
Want a second start?"

"You bet, Dear Lord, my mind's made up.
I've made my final choice
Yes, I'll be your brand-new son!"
Let angels now rejoice!

"Partake, Joint Heir, your heritage . . .
I am your great Reward
Stick closely by my side, Dear Child
I'll guard you with my sword!"

Friend, are you wandering lost about,
An orphan of the soul,
Bleeding from your brokenness?
But One can make you whole.

Cease waiting 'til you're good enough
There's nothing you can do
God's business is adoption, Child,
And He has chosen you.

Forsake the earthly vanities
No treasure's left to own
That equals that sweet moment when
God says, "Child, welcome home."

Panic

Within just a few days I saw signs obvious to any mother that Michael had been traumatized. He did not express the normal emotions of a four-year-old. No matter how badly he was hurt, he did not cry. It was little wonder. Even the few facts we knew would have been enough to scar a child for a lifetime without the mercy and intervention of God. Tears had not helped him much through neglect, continuous abandonment, constant conflict, broken promises, harsh punishment, and an early introduction to the police and Child Protective Services. He was in emotional shambles.

Not only did he seem void of tears; he was also void of laughter. No matter how silly his new sisters acted, he rarely smiled. He had no appetite except for things, and there were never enough of them to satisfy his emptiness. He entered our home on the basis of one assumption—sooner or later we would leave him, too, and he'd just as soon cut to the chase. It seemed to become his goal in life to hasten the inevitable.

He required the focus of all attention and within a very short time was terribly threatened by the two children who had preceded him there. If either of our daughters came to me for attention, he would find a very effective way to seize it for himself. At the same time he craved my focus, he also remained strangely detached. He seemed to have very little feeling for me until nighttime when he would hold my head in a death grip and drop off to sleep chanting, "Please, Mommy, don't leave me."

Michael suffered constant bouts of deep depression that would ultimately turn to fits of violence and anger. Emotion finally emerged with frightening velocity. Our neighbors would not let their children play with him. His preschool teachers could not deal with him. We tucked our tails and ran as fast as we could to a godly child psychologist who met with him four times. The fifth appointment she found me in the waiting room instead of Michael. In near hysteria, I cried, "Help me. Teach me. He needs a counselor twenty-four hours a day, and I'm the only one there."

By this time our other children were traumatized too. Like their mother, they had envisioned the adoption to be a glorious romance and found themselves, instead, being displaced by a child who was ready to seize attention at all costs. They had tried their hardest to reach out to him. One day Amanda was trying to get Michael to draw. We had often tried to get him to express himself on paper only to see him scribble angrily until the paper was torn to shreds. We were too ignorant to realize he was expressing himself.

This particular day she changed the proposal to one that would meet with his approval. "Michael, look at the stick man Sissy drew." He glanced at her drawing and cracked a tiny smile. She continued, "If I draw you a stick man, will you just fill in the face? That's all. Just draw eyes and a nose and a mouth. I know you can do it. Show me how." To our amazement, he nodded his head, and we all stood around him with great anticipation. This would be the first picture Michael had ever drawn in his life. The face he drew on that stick man, he also drew on my mind forever. The mouth was turned down and lines of tears streamed from the eyes. We all sat and stared. Amanda had the maturity to say, however sadly, "That's really good, Spud. Really good. Thank you." I excused myself from the table, went into my room, put my head under my pillow and sobbed. At that very moment I knew that precious child needed more than we had. Why hadn't God given him to parents who really knew what they were doing? Who didn't have other children? Who didn't have such demanding lives already? This one was beyond me. It was the most terrifying season of my adult life.

Could It Be?

Empty eyes, a fragile heart
Where to begin? How do I start?
Consider the risk! The pain's too intense!
I can't be the one. This doesn't make sense.

Even his sleep isn't peaceful or sound
From a crack in the door light casts on a frown.
He's made up his mind by the time he is four
Too much behind him . . . too little in store.

A smile so rare, his words so few
I study expressions, they offer no clue
No laugh when he's tickled, no cry when he's hurt
Won't run ahead nor hide in my skirt.

Desperate for playmates but unable to play
A few moments pass, they all run away
Surrounded by people, he's still all alone
No sense of family, no sense of home.

In one breath he utters, "Please, Mommy, don't leave."
Then he shoves me away, my heart starts to bleed
The walls go up, he slams the door
Never wanted me less, never needed me more.

Can't you see I'm torn to shreds?
Precious lives at stake in nearby beds

Courage melting, what have we done?
Satan fights dirty . . . sometimes he's won.

So many questions shout in my mind
You make no mistakes but maybe this time
You entrusted too much and caught us off guard
Or could it be
Your perfect will
Is sometimes just this hard?

Fresh out of methods, struck out on my plans
Nowhere to quit . . . can't wash my hands
Filled with self-hatred, hanging my head
God lifted my chin then gently said,
"But,
Can you just love him?"

Questions

I'll never forget the moment I knew God was asking me that simple question, "Can you just love him?" It was the night I had come to the end of my strength. Keith and I had tried to get away for a weekend to catch our breath. My parents had offered to babysit our boy, and it had been disastrous. Keith's parents were equally concerned for us, and our friends wondered what we had done. We felt as if we were sitting all alone in a sinking boat. After he had gone to sleep that night, the rest of us sat on our oldest daughter's bed, held one another and cried. Our daughters were growing resentful and felt like they couldn't get to us any more. We were stretched to the point of ripping.

Later that night, when we were all alone, I laid down the law to Keith. "If we happen to be strong enough to pull him up, he stays. But if he is strong enough to pull us down, he goes. I will not have my other children traumatized because of a child who refuses to be helped." Keith stood up and walked out the front door. He looked up and down the street then walked back into the house. "What are you doing?" I asked angrily.

"I'm looking for the next set of parents in line, and I don't see a single volunteer. Elizabeth, this child has been abandoned over and over again. We told him we wouldn't be like them. This is his last stop."

I burst into tears, "Then what are we going to do?"

Before Keith could say a word, I heard it. The voice of God speaking loudly . . . speaking directly into the depths of my heart. "Can you just love him?"

He could have asked me a multitude of questions I could have readily answered—

"Can you make sure he's educated at the best schools?"

"Yes, Sir, I can."

"Can you always dress him to look his best?"

"You bet I can, Sir! He'll be the best dressed in his class!"

"Can you take him to church?"

"Without a doubt, I can!"

"How about Disney World?"

"Consider it done, Sir!"

But that's not what He asked.

"Can you just love him? Really love him? With a love that never fails?"

Before I knew what had happened, Keith was pulling me up the stairs saying, "We're going to pray over him, that's what!" I practically crawled up those stairs, totally exhausted physically and emotionally. We asked God to leave Michael in a deep sleep as He had Abram when He made a covenant over him. The covenant we needed to make was between God and the two of us. We placed our hands on that sleeping boy and mustered up the little strength we had left and cried out for help. We admitted to our utter helplessness, our worthless naiveté, and our extreme disappointment. We claimed that boy for Christ that day like we never had before, and we invited the enemy who had slipped in our house through a crack in the foundation out of our home. We walked down those stairs feeling centuries old.

It wasn't the next day. It was so gradual I couldn't even tell you exactly when it was. But one day things began to change. In bits and pieces. In glimpses and glances. Somewhere along the way, we became his parents.

From Where I Am

Three years have come, three years have gone
Sometimes it seems short, sometimes it seems long
Never had so much doubt, never made such mistakes
Nor had so much worry nor kept so awake.

Three steps forward and two steps back
A life full of grays . . . I like white and black
Don't know how he seems from up where You are
But this is the guy I see in my yard—

A handfull of frogs, a hose in the other
He struts to the meter to flush out another
Eyes once so empty, now full of mischief
Jean pockets bulging with all kinds of kid's stuff.

One shoe on the porch the other is missing
Ears prop his cap, they're sure not for listening
He'll fish at the pond if time will afford
With a baseball nearby in case he gets bored.

His room is cluttered with artwork from school
He likes hightops and haircuts that make him look cool
He'd gladly dump school, thanks all the same,
'Til he walks down the hall, and kids call him by name.

Just when I've lost it, frazzled and frantic
He swipes flowers from the neighbor's,
he's such a romantic
Paddled a plenty, his feet stomp the floor
I'll think that it's settled, he'll try me once more.

We've read lots of books and fought lots of fears
We've talked our mouths dry and shed lots of tears
"Mommy, did you miss me before I moved in?"
And, "I want to be born from your tummy like them!"

We've come a long way with a long way to go
No clues for the future, so much I don't know
Will he ever trust You to set his soul free?
Or realize the teacher that he's been to me?

I'm not all alone where answers are due
I recall, there's an answer I still owe to You
Three years have vanished, It's time to confess You asked,
"Can you just love him?"

Yes, Lord, oh yes.

Love

The Greeks called it *agape*. God called it a supernatural expression and something only He can do. But He does it through hearts vacated by their own responses and made available for His. Agape is a kind of love God demonstrates to one person through another. Romans 5:8 gives us the perfect example, "But God demonstrates his own love [agape] for us in this: While we were yet sinners, Christ died for us."

In other words, God loved us through Christ. His Word also says, "As I have loved you, so you must love one another" (John 13:34). Just as He demonstrated His love to us through Christ, God desires to demonstrate His love to others through us.

No doubt, God has placed you in the position to love the unlovely, whether or not you've been obedient to that command. In fact, I think that we can surmise that anyone among us who is not struggling with loving someone isn't getting out enough! It's one of His priority agendas. He wants to use us to love with a supernatural supply someone we never could love otherwise. No believer can avoid this crucial high calling. Agape is an obedient response of availability to God, not a feeling. But, although it is not a feeling, its ultimate end is a feeling. You'll never respond to God with the mind of Christ that you do not finally end up with the heart of Christ. It's nothing less than supernatural.

A pure, unadulterated miracle. See it for yourself—"Love never fails" (1 Cor. 13:8).

You may ask, "Does it never fail the giver or the receiver?" It fails neither. For the receiver, they will never have been loved for nothing. God is very practical. If He has called upon you to be His vessel of love toward someone else, it is because He has a plan. If you are obedient, the working of that plan is God's responsibility and not your own. In other words, Keith and I have been called to be instruments of God's precious love to Michael. It is our responsibility to be obedient to God through availability. The final results of that love are between God and Michael. Keith and I are hoping that the results will be a young man totally surrendered to God and in love with His Savior, but even if he never accepts the love of God he has been shown, we must not fail to show it.

Love also never fails the Giver. We've learned through firsthand experience that when you agree to let God love the unlovely through you, He never fails to make the unlovely lovely to you. Michael resembles nothing of the child who came to us that unforgettable Valentine's Day.

Did the child ever learn to cry? Louder than anyone you've ever heard. The sort of weeping, wailing, and gnashing of teeth that would make an ancient Israelite proud. And, boy, did he learn to laugh. One day when he was in the first grade, I received a call from the principal's office at his school. (I'm quite sure we're on speed dial at this point.) It seems that he walked into the boys' restroom during lunch while three little kindergartners were lined up at the stalls. This particular restroom had no windows in it, so when he got the bright idea to turn off the lights, they reacted by screaming and turning round and round in a frenzy, thus effectively spraying the walls and one another. Michael was discovered minutes later still folded up on the restroom floor holding his knees to his stomach laughing until he was in pain. As I understand it, he was carried in that very posture to the principal's office.

As you can obviously see, Michael is no longer painfully detached. He holds his own in the midst of a very active family. He vies for his rights but not nearly so often at the exclusion of another's. He adores his sisters, and they adore him, and they all seem to enjoy annoying one another to pieces. He tells stories with enough animation to make Walt Disney rise from the dead. And remember that tragic picture he first drew? Just one year later his kindergarten drawing of a Texas Longhorn won a district blue ribbon to pin on his boastful chest. He's miraculous, mischievous, and marvelously Moore. He still forces his teachers to earn their paychecks and his parents to stay on their knees. He is far more likely to get in trouble for entertaining other children than hurting them.

God has taught me things through Michael that I never learned in that child development course all those years ago. Three stick out in my mind above all others—

> 1. Perfect parents don't exist, but a perfect God does.
>
> 2. Agape is hard work. But it always works. It may not always have the results I want it to, but it will always have results.
>
> 3. Sometimes agape really hurts. It broke the heart of God to demonstrate His love to us through Christ but its ultimate end was salvation.

There will be times that it will break our hearts to be vessels of God's love toward another, but its ultimate end is meant for salvation. The salvation of someone's soul, health, reputation, marriage, honor, sanity. Through love He saves.

To the grace and glory of our Rescuer Up Over, I believe that our son's life is good. Not easy but good. Michael has been forced to work harder at life in his few years than many others will in a lifetime. How hard he has had to try just to be normal may not mean much to others, but to a set of parents who knelt in desperation over a young child's bed not many years before, it means everything. When the ultimate "Award's Day" arrives, these feats will not go unnoticed.

Award's Day

I went to my son's school that day
It was a very special day
When worthy tribute would be paid
To honor students in first grade.

Music ushered children in
Faces wet with toothless grins
Flags were raised and banners hung
Pledges said and anthems sung.

I stood with other moms in back
He didn't know I'd come, in fact
I didn't want his hopes set high
In case his teacher passed him by.

Every mom felt just the same
All had come to hear one name
The child she hoped they'd recognize
And find deserving of a prize.

The list went on page after page
As beaming children walked the stage
Cameras flashed and parents cheered
Grandma smiled ear to ear.

My eyes were fastened to just one
The anxious posture of my son

Perched at the very edge of seat
Too young to have assumed defeat.

Certificates for everything
From grades they made to how they sing
For days not missed, for how they drew,
Good citizens to name a few.

But it wasn't likely on that day
They'd honor one who'd learned to play
And stay in class from eight to three
Who'd learned to write and learned to read.

We hadn't hoped he'd be the best
We prayed he'd fit in with the rest
I knew no matter who they'd call
My boy had worked hardest of all.

An elbow nudged me in the side
A friend attempting to confide
A boy waving frantically,
"There's my mom! Right there! You see?"

They never called his name that day
I drove straight home, sobbed all the way
The boy? He had ceased to care.
He had a Mom and she was there.

Loss

One of the greatest difficulties of Christ's earthly experience had to have been "knowing all that was going to happen to him" (John 18:4). He came to His own nation knowing they would reject Him. He loved Judas knowing he would betray Him. He spent intense hours with the disciples prior to His trial knowing they would forsake Him. Yet the depth of Christ's love and His willingness to face the Cross remained unchanged. In God's infinite wisdom, He chose for us to remain completely unaware of the experiences awaiting us. Sometimes we say, "If only I had known . . . " Far more often, we see the wisdom in not knowing.

When I wrote those last words about our beloved Michael, I had no idea he would one day return to his birth mother. As he approached preadolescence, serious internal struggles began to surface in the form of alarming behaviors. Although we never doubted Michael's love for us, he seemed unable to control many of his actions. In his childlike way, he was spinning in the cycle so aptly described by the apostle Paul. "I do not understand what I do. For what I want to do I do not do, but what I hate I do" (Rom. 7:15). Keith and I sought the expertise of countless doctors, specialists, and Christian counselors. We transferred Michael to one of the finest specialized schools in Houston and wore our knees to the bone, praying through oceans of tears. We were told over and over that Michael had needs beyond those we could meet. We asked God to confirm to us whether this counsel was true. We desired with all our hearts to keep Michael forever but knew if we were no longer helping him, we must find someone who could. To our shock and utter dismay, God confirmed He had a plan necessitating a drastic change.

Through circumstances so "coincidental" they had to be ordained by God, Michael's birth mother resurfaced, strongly desiring to reclaim her son. She was trying to put the pieces of her life together for the first time and believed she was ready to try to be a parent. Eight years earlier, I sat across a table from her in a restaurant and asked, "Are you sure you want us to have Michael? If you want to try to raise him yourself, I could simply help you." She admitted she was in no position to raise a child and granted him to us. As we drove away from that restaurant, I cried to my husband, "She needs a mom as badly as Michael does."

She remembered our conversation. When she resurfaced, she said, "I want to raise him myself. Would you help me now?" Her name is Anne. She is not only Michael's natural mother but also a close relative. We could not accept this sudden twist as simple coincidence. We did not make the decision to allow Michael to return to her for her sake. After extreme prayer and deliberation, we consented for Michael to live with her at least for a while for his sake. We pray and believe she holds an important key God wants to use in his healing.

I have come to a startling conclusion. In Christ we are capable of things we never dreamed. We have entrusted our boy to his birth mother's primary care. Through grace and strength which could only come from Heaven, I am attempting to do what Anne asked, to help her learn to be a mother to Michael. We are two women with absolutely nothing in common . . . except for one little boy we both call Son. We have a strange bond, and actually, a peculiar love for each other. We recognize his need for both of us and are both learning to share. I have never before experienced this kind of stabbing pain and loss. Words defy an explanation of the emotions my family has experienced.

We are surviving our loss on one solitary basis: we believe this is presently God's will. We wanted God to heal Michael in our home. However, God sees a grander picture. Another life is at stake here. Anne is learning to trust Christ for the first time in her life. She quickly confronted many difficulties but remains steadfast in her desire to do everything she can for Michael. I am proud of her. We have no idea what the future holds or how much or how long she will be able to help. We have learned hard lessons about presuming permanence in terms of God's will. We know now that God always intended for Michael to be with us for a season rather than permanently. In my agony I cried out to God, "Why is this happening?" His answer came clearly to my heart. "For six years you have worked, and on the seventh year you will rest." I was totally stunned. I realized God had planned this all along, waiting for a certain troubled birth mother to come knocking on His door. She came just shy of seven years. Our hearts are still in so much agony but our spirits are at rest. We believe we are in God's will.

I am so thankful I did not "know all things that were going to happen." I had no idea I would ever lose Michael, so I loved him without restraint. I never dreamed he would return to his birth mother, so I didn't love him like an aunt. I loved him like a mother. I fell short in so many ways. I wish I could have dozens of "do-overs," but I gave him everything I had. I will love Michael like a son until the day I die. We still see each other periodically, and we hold on for dear life. We laugh and cry and try to understand. God has been faithful. He holds out healing to anyone who will open his arms and receive. May all who are involved in this difficult situation accept His healing. But above all, may one beautiful little boy named Michael call upon the name of the Lord and be whole.

The Life I Planned

Has someone seen the life I planned?
It seems it's been misplaced
I've looked in every corner
It's lost without a trace.

I've found one I don't recognize
Things missing that were dear
Promises I'd hoped to keep
And dreams I'd dreamed aren't here.

Faces I had planned to see
Hands I planned to hold
Now absent in the pictures
Not the way I told.

Has someone seen the life I planned?
Did it get thown away?
God took my hand from searching
Then I heard Him say,

"Child, your ears have never heard
Your eyes have never seen
Eternal plans I have for you
Are more than you could dream.

"You long to walk by sight
But I'm teaching eyes to see.
I know what I am doing
'Til then, you must believe."

He's done so much, I felt ashamed
To know He heard my moans
To think I'd trade in all He's done
For plans made on my own.

I wept over His faithfulness
And how He'd proved Himself
How He'd gone beyond my dreams
And said to Him myself,

"No, my ears have never heard
My eyes have never seen
Eternal plans you have for me
Are more than I could dream.

"Yes, I long to walk by sight
But You're teaching eyes to see
You know what You are doing
'Til then, I must believe."

I felt His great compassion
Mercy unrestrained
He let me mourn my losses
And showed to me my gains.

I offered Him my future
And released to Him my past

I traded in my dreams
For a plan He said would last.

I get no glimpse ahead
No certainties at all
Except the presence of the One
Who will not let me fall.

Are you also searching
For a life you planned yourself?
Have you looked in every corner?
Have you checked on every shelf?

Child, your ears have never heard
Your eyes have never seen
Eternal plans He has for you
Are more than you could dream.

Perhaps you long to walk by faith
But He's teaching eyes to see
He knows what He is doing
Child, step out and believe.

"No eye has seen, no ear has heard, no mind
has conceived what God has prepared
for those who love him."
1 Corinthians 2:9

Cost

These words are for anyone who has ever occasionally counted the cost of her calling by the drops of her tears. One of the high costs of my calling is God's requirement upon me to kiss my children good-bye and "go ye therefore." Every couple of weeks, when I board another airplane, no matter how brief the excursion, I must trust my God to explain the "what for" of their mother's "therefore." I've found Him faithful. So far it has seemed that He is never more clearly there than when I am not.

Homesick

I called to check on home last night
To see if all was going right
My man assured me all was well
And it was true . . . I could tell.

I felt so far away from home
So by myself, so all alone
No noise here, no bouncing balls
No fussing kids, no endless calls.

I asked if everything was set
I didn't want him to forget
To take care of the "mother things"
To hang their shirts and crease their jeans.

He said, "Your oldest set her clock.
She'll get us up right on the dot
Don't worry, now, they'll get to school
We love you much, we'll see you soon!"

The phone went dead. I wasn't through . . .
I barely said, "I love you, too."
I sat and stared down at the floor
"She's never set her clock before."

She's just a kid, not old enough
To wake without a mother's touch

What chance is there at school they'll say,
"You're one great kid! You're loved today!"

Kids need to hear those words first thing
Before a careless clock can ring
And furthermore, they like, I frowned,
Hot cocoa when they first come down!

"Dads," I thought, and fell in bed
Then after while, to myself said,
"He's probably right, give them a break
She is fifteen, for heaven's sake."

"Fifteen," I sighed, "Where has it gone?
Since that first day before the dawn
When she and I told secrets dear
And her first bath was in my tears?"

I'd held her close with just one arm
Reached for the phone to call my mom
"Oh, Mom," I sobbed, "I love her so!"
She cried as well and said, "I know."

The years are mean . . . they rush on by
The kite string slips into the sky
She's nearly grown, yes, plenty old
To wake up when the clock says so.

I felt so sudden like a fool
It won't take Mom to get to school
How silly . . . they will all be fine
Just go to sleep and rest your mind!

I tried to let the dawn go by
Without a call to check and pry
To see how every one had fared
Got your lunch? Homework prepared?

I finally grabbed the phone and dialed
It seemed to ring a country mile
My heart sunk swift . . . they must be gone
Dad's out the door . . . dog's on the lawn.

I started to hang up the phone
Until I heard a voice on
The other end, as up he leapt
"For heaven's sake, we've overslept!"

Suddenly the house lit up
He threw the phone, said, "Kids, get up!"
I heard each voice at a time
They were mad, but they were mine!

I cheered them on from miles away
I heard them readied for their day
And just before they slammed the door
She yelled, "Thanks, Mom!"

That's what I'm for.

FRAGILE PLACES

Memories

There are times God meets us on our journey with others just as He did in my marriage, the birth of our daughters, and the adoption of our son. There are other times when God reserves the right to meet with us all alone. These are the times we search our personal worlds with desperation and find ourselves lacking. Times when closest of families cannot help and close enough friends cannot be found. Times when God Himself busies telephone lines, deafens ears, and blocks understanding. Times when He chooses not to reveal Himself through any living flesh but commands that you learn to look straight into His invisible face. These are times when we journey to the extremity. Times when answers never come nor would the end lest our "why's" take us there and our "who" brings us home. It's a place where we can only go alone. And find Christ alone.

My unscheduled journey to the extremity accompanied the restoration of traumatic childhood memories. Someone who should have loved me used me . . . someone who was a traitor to my parents' trust. Unable to process the memories until adulthood, the void had been filled with an unexplainable sense of shame and an insurmountable search for safety. Never delude yourself into thinking Satan does not prey on the lives of children. His wickedness has no bounds nor is any earthly ground sacred. He seeks to disqualify us when we're old by corrupting us while we're young. Often we try desperately to determine whether pain has come to us from the Throne of God or the Kingdom of Darkness. There is one sure way to know when Satan has been at work. SHAME IS THE SIGNATURE OF SATAN. As it was in the Garden, so will it ever be. Shame is Satan's "been there, done that." With shame comes the inevitable prison of secrecy. With secrecy comes loneliness. With loneliness comes poor choices of company. With poor choices come more shame. More defeat. More of the enemy.

The strength of the chain intensifies and chokes abundant life. God has taught me that the secret to living is not living in secret. For there is where the shame is often bound. He used the recovery of painful memories and the reclaiming of my past, not to bind me but to set me free. I arrive at this point in my journey void of the words you may be accustomed to hearing. Contrary to the claims of others, I am not glad it happened no matter how much I have learned. Only the foolish could make such judgments concerning crimes against a child. Certainly Christ would have no such expectation. Hebrews 12:2 tells me that He "despised the shame"(KJV, author's paraphrase). I cannot fellowship in His sufferings without also despising the shame. But is also says "for the joy set before him" He endured. I have indeed discovered a taste of that joy, and He was worth the inevitable journey. There is quite a difference between being glad something tragic happened and being glad something happened out of the tragic. Through its lengthy course, when I could walk no more, I rested on the side of the road, rubbed my swollen feet, and poured out my aching heart. These are just a few stops along my healing way.

Gently, Lord

Love me gently, Lord
I'm hurting now.
I've lived to see Your sovereignty
You've taught my knees to bow
I've caught glimpses of Your glory
I've seen Your righteous ways
But right now I need You, Father,
Just to face another day.

You have promised not to always be
Exactly what I please
But You give me sweet assurance
You're exactly what I need.
I need a gentle Father
And the lullaby He sings,
"Let Me tuck you safely
Underneath My healing wings."

Love me gently, Lord,
I'm hurting now.
You said, "Take your cross and follow Me."
I beg, please show me how
To celebrate my weakness
That in You I might be strong.

When desperation grips my soul
A moment seems too long.

Oh, God, what noble plans I had
To do this whole thing right
Now I fall before You wounded
And I've lost the will to fight.
There are soldiers all around me
They're depending on me, too.
I fear I've nothing left to give
So, again I ask, Can You?

I'll love you gently, He says,
I know you're hurting now.
You've oft revered my sovereignty
Your knees have dropped to bow.
If you could only see things
From My throne's clear point of view
You'd see glimpses of My glory
Are fast at work in you.

I'll love you gently.
Let Me soothe your hurting now.
I've said, Pick up and follow—
I'll do more than show you how.
I'll turn this Throne of brilliance
Into a rocking chair.
Crawl aboard, My precious child,
And I will rock you there.

Hide

Hide, little girl, beneath your Mama's skirt
Hide, little girl, and maybe it won't hurt
Hide from the laughter, what if it's at you?
Hide from the sorrow so no one has a clue.

Hide, little girl, behind the smile you learned
Hide beneath the masquerade of credits that you earned
Hide in crowded corridors until the school day ends
Hide in courts of favor but never trust a friend.

Hide, little girl, behind your wedding veil
Eyes that cannot cry are eyes that tattletale.
Hide, little girl, until your time runs out
Can't always hide, little girl, one day you'll be found out.

You can run, little girl, to the only One who knows
To a place of fertile soil where trust can finally grow
Then you can hide, little girl, 'til every eye may see
You found, little girl, safely hid in Me.

"Hide me in the shadow of your wings . . .
that I may gain Christ and be found in him."
Psalm 17:8; Phillipians 3:8–9

Tricycle Dreams

Some dream of greatness,
    others of fortune.
        I dream of innocence.

And innocence has a face . . .
    cheeks pink with sunshine,
        eyes bright with hope,
            lips wide with laughter,

Propped carefree on the seat of a sailing trike,
    Curls bounce with the rhythm of percussions on
        parade.

Bare feet turn the pedals as she spies her destination,
    a trail of dust nipping at her heels.

No need to look back . . . there's nothing behind her.
    No past. She's only four.

She thinks tiny girl thoughts and makes tiny girl choices
    like how to dress up her doll.

Mama's best heels lay strewn in the grass,
    her pearls drape a tiny girl neck.

She likes Red Rover . . .
    likes for kids to come over . . .
        She'd rather go seek than hide.

She'll change her mind again tomorrow
    about who she wants to be
when three small wheels can no longer take her
    where she'd please.

And one day pain will surely come
    but it will greet her just as it should . . .
        by surprise.

Till then, fairy tales will consume her thoughts
    and trust will blind her childish eyes.

Yes, some will dissuade me and think it a waste
    and far too late for tiny girl dreams,

But if through my prayers
    Christ gently spares
        One of these little ones,

Then dreams come true
    And there is innocence.

        Enough for me to share.

Extremities

Satisfy me not with the lesser of you
Find me no solace in shadows of the True
No ordinary measure of extraordinary means
The depth, the length, and breadth of You
And nothing in between.

Etch these words upon my heart knowing all the while
No ordinary roadblocks plague extraordinary miles
Your power as my portion, Your glory as my fare
Take me to extremities,
But meet me fully there.

FINER
GRACES

Changes

I was an accident waiting to happen. The only one of my parents' children born without the benefit of anesthesia in the days when most mothers preferred the "wake me when it's over" method of childbirth. My very funny, ever so slightly eccentric mother reminds me quite often just how much I hurt. She calls me every morning on my birthday to say she has the cramps. By noon I've received some sort of hate mail from the postman. One year it was a card printed in Spanish. The only reason I pursued a B.S. degree in college rather than a B.A. was to avoid foreign language. Needless to say, I couldn't read a word of it, and neither could she. In fact, it probably wasn't even a birthday card. The bottom was simply signed, "Love, Mom." Another year she sent me an ancient Christmas card she had found in an old box addressed to her from my Aunt Irene. It was yellow, brittle, and ugly. She had marked through the words "Merry Christmas" and written "Happy Birthday." She even left my Aunt Irene's note at the bottom of the card, then marked out her signature and inserted "Mom." When all of us are at my parents' and someone suggests the kitchen needs cleaning, she's been known to remark, "Let Beth do it. She hurt the worst." And that was only the beginning.

Within a few minutes of my birth, I had an allergic reaction to the eyedrops the doctor administered, and I formed blood clots on the surfaces of both eyes. Not one of my siblings nor I had any hair until we started school, and my skin was so pale and translucent you could see every blood vessel. Bloodshot eyes, a bald head and see-through skin. I looked like a shiny road map. I was lamenting my rough start to my Grandmother once, and she replied, "Yep, people used to come by the house and peak into your bassinet, and all they could think up to say was, 'Isn't she young, though!' "

About the time my eyes began to clear, the army pediatrician pronounced me hopelessly pigeon-toed unless I succumbed to years of corrective shoes. They were the first things I put on in the morning and the last things to drop with a thud from the featherbed I shared with my grandmother at night. They were hideous. The soles were an inch thick. The colors were black and brown. When I grew out of one pair, I grew into another just like them. I longed for black patent leathers and barefoot sandals. Better yet, I wanted to be barefooted . . . a proud Arkansan's inalienable right. I wore those wicked things until I went into the first grade, long enough to have developed a lifelong shoe fetish. To this day, when I walk into my closet, I want to see shoes. I don't care what color, what year, or what model. I want to see shoes. I don't care if they're not my size. I don't care if they're not mine! I want shoes!

The very week I got to shelve my corrective shoes, my older brother, whom I still adore, walked through the door and held his hands out to me. That was my invitation to go flying across that floor into his arms. And that is exactly what I did. Someone had moved a piece of furniture just enough to place a very expensive fold in our oval cord rug.

With my freshly straightened feet, I tripped and went air-borne. I landed mouth first on the edge of the coffee table, shoving my top baby teeth into the roof of my mouth. In no time at all, my teeth had turned black from the roots I had inadvertently murdered. I couldn't wait to get those horrible teeth out of my mouth so I could look "pretty" again. I knew it couldn't be long because I was at that age when children lose more teeth than their mothers do car keys. I jerked and jerked on them, and, finally, one by one, out they came. I couldn't wait to grow in my new pearly whites. I woke up every morning and inventoried my top gums to see if there was any sign of them. After endless disappointing mornings, I awakened to a very strange sensation in my mouth. I just knew this had to be it. I immediately felt my gums and dis-covered nothing. I knew I felt something, so I jumped up, ran to the mirror, and lifted up my top lip. It was them all right. My brand new teeth. Growing straight out of my top gums. No, I don't mean buckteeth. I mean teeth you could play chess on. Teeth that prevented my lips from touching for years to come. Teeth covered with some form of metal for twelve solid years. I feel sure I could qualify for the Guinness Book of World Records but who wants to be known for hav-ing the worst case of bucked teeth in the history of modern man? I had enough problems. The kids at school made fun of me, and I had a heck of a time drinking my milk through a straw.

Just about the time I entered the fifth grade, my teeth were beginning to look somewhat southward. Then a hor-rible thing happened. It was something I had dreaded all my life. It had happened to all of my older brothers and sisters one by one. It was the generational curse. I thought fate might be kinder to me since, after all, buckteeth should be the only suffering a person should ever need to mold their character. I was wrong. It's a wonder I didn't feel it happen. The night before it wasn't even there. I'm quite sure my head was a little heavier to pick up off the pillow, but I've suffered memory loss over the trauma. I staggered to the mirror to brush my bushy hair and get ready for school and screamed. There it was. The family nose. And it was a doozy. My face practically had stretch marks. My eyes looked suspiciously closer together. It was bad, all right. It came from my Daddy's side of the family. The only good thing to be said about it is that it gets a little smaller with every generation. You should see his. As soon as he opens his car door at Kroger, the automatic doors open. They say his Dad had to be buried with an open casket.

In no time at all, "fishnet" hose became the latest rage. They came in every conceivable color, and, boy, were they cool. Goodness knows, I could've used a little cool, so I saved my money and stocked up on every shade. Luckily, my mom was not like her stuffy friends. She cheerfully allowed me to make my fashion statement. There was just one catch—she wouldn't allow me to shave my legs. And did I ever need to shave my legs. The hair on my legs was at least an inch long and when I pulled up my fishnets, which took no small effort, I looked like an eighty-pound porcupine in a chartreuse hairnet.

By the time I was sixteen, I had been fired from my first job—assistant peanut-butter grinder at The Nut Hut. I accidentally ground my thumbnail into a patron's peanut butter.

I might have gotten away with it, but my nails were painted red. Apparently my boss did not consider a broken nail punishment enough. These days a woman could sue over a thing like that.

I needed the Lord. In the midst of multitalented brothers and sisters who had all somehow discovered their niche, no matter what my parents said, I felt ordinary and insignificant. Only one thing seemed to challenge my assessment. That man in those watercolor pictures my Sunday School teachers continually talked about. The One who calmed a sea, and more impressively a boat full of scaredy-cats. The One who thought women were as neat as men. The One who surrounded Himself with children. The One I had accepted as Savior when I was a small child and found myself falling in love with as a young woman. No earth-trembling testimony. No tale worth retelling. I met Him at home and fell in love with Him in Sunday School. For a little Cinderella who couldn't find her Fairy Godmother, He was my hero. My Knight.

As the years passed, things changed. I was no longer a nerd. I had grown into my nose a bit. (Trust me. I still show signs of nasal trauma.) Rollers calmed my hair. I was named everything from sorority president and All-Campus Favorite to Who's Who in American Colleges and Universities. Everything had changed. Except one.

> Jesus Christ . . . the same yesterday and today and forever.
>
> HEBREWS 13:8

I hadn't believed much of my own press. I still don't. Way down, deep inside is that same pigeon-toed, bucked-toothed little girl. And way up at my Father's side is that same Hero that touched the leper's spots and made him whole. And one day . . .

"After my skin has been destroyed, yet in my flesh I will see God. I myself will see him with my own eyes . . . [on] a white horse whose rider is called Faithful and True. . . . How my heart yearns within me!"

Job 19:26–27; Revelation 19:11; Job 19:27

Heroes

In a world of philosophers, amateur and not
Exists arrival at a venue where agreement may be sought—
There are no heroes.

The young are oblivious to something worth defending
They rally 'round the reason on which they're all depending—
There are no heroes.

Statistics boast the aging are aging further still
While disease will never match the number emptiness will
    kill.
There are no heroes.

Where is the one 'round lines are formed
    for autographs worth sharing?
Who has a name proud parents steal and want their new-
    born bearing?
There are no heroes.

In a cold wet corridor through which a subway thunders
A young graffiti artist takes spray paint and wonders—
There are no heroes.

As lack of hope takes prisoners, if the soul within you shrinks
Let history bear her testament when you begin to think
There are no heroes.

To a man without a country, He appeared a joint sojourner.
To Joshua armed but afraid, He came a valiant warrior.

To Moses raised up on the mount, He was the One yet higher.
To Shadrach, Meshach, and Abednego,
	He was the fourth man in the fire.

To Elijah who stood as one for God, he was never less alone.
For Noah's faithful family, He made an ark their home.

To Ezekiel He appeared to be the light cast o'er the dark.
To King David running from the throne, He was the true
	Monarch.
To Daniel at the bite of death, He was the lock upon their
	jaws.
To King Solomon who'd had it all, He was the only worthy
	cause.

To a sinking fisherman, He was life upon the water.
To a grieving Jairus, He was life unto his daughter.

To a woman at the well, He was complete acceptance.
To a doubting Thomas, He was the proof for his reluctance.

To a dozen throwbacks from the world,
	He unleashed His awesome power.
From a greedy grave of several days burst forth his finest
	hour.

You may reply, "I don't accept such stark mythology
Written much too long ago—what have they to do with
	me?"

My friend, I beg you listen for in each there is a clue
What He has done for all of them He waits to do for you.

If you thirst, I know some flowing water live and able.
If you hunger deeply, He invites you to His table.

If your eyes are blinded, He's the One to make you see.
If you're in life's strongholds, He's majestic liberty.

If you're lost, He is the only Truth, the Life and Way.
If there is a debt you owe, there is but One who'll pay.

He is the Christ, Redeemer, the Son of God, adore Him!
He is the matchless, Holy One, and all will bow before Him!

So you may scale the mountains high
        and search the valley low,
But until you meet my Jesus, you'll search in vain,
He is the true Hero.

A Matter of the Heart

I've arrived at a conclusion,
Maybe one of life's rare finds
That there's not a lot worth salvaging
Within this heart of mine.

It's ever ready to destruct
And lie above all things . . .
It tends to laugh when it should cry
And mourn when it should sing.

I've wasted countless hours begging,
"Fix this heart, Lord, please!"
While it stomps its feet, demands its way
And floods with sin's disease.

At last, You're able to get through
And lay it on the line—
"You must give up that heart of yours
And trade it in for Mine."

So I cry out with the psalmist,
Create within me, Lord
A new heart crystal clear
That only Calvary could afford.

A heart which pounds the rhythm
Of Heaven's metronome
And issues forth a boundless love
And beats for You alone.

I want to love that which You love,
Despising what You hate
And see myself as least of these
Oh, Lord, retaliate.

The efforts of the evil one
Who seeks to make my plea
That of his own, "I'll make no move
'Til I've considered me."

Peel away my fingers,
Finally make me understand
The power to love and please You
Can't be found within a man.

So, my Lord, I bring this offering—
A stubborn heart of stone
And ask You, in its absence,
Please exchange it for Your own.

How

How would I have known that I was lost
Had You not searched and found me?
How would I have known that I was blind
Had You not made me see?
How would I have known my bleeding
'Til You bound Your love around me?
How would I have groaned my slavery
Until You set me free?

Author, Finisher of my faith,
It all begins with You.
I'd still be wandering in the dark
Alone without a clue
That somewhere beyond the fairy tales
A child's dreams come true
Every time she risks her heart
And rediscovers You.

The Day He Called My Name

Schedules to keep, people to meet,
  it's a wonder I heard You at all.

Over city screams and midday dreams
  prevailed Your whispering call.

For only a moment, too brief a moment,
  You beckoned me up that hill.

Engagements were broken, apologies spoken,
  "I've a divine appointment to fill!"

I grabbed a canteen, threw a pack on my back,
  fully stuffed with all I hold dear.

Then I heard You reply with a smile and a sigh,
  "I don't think you'll be needing those here."

Piece by piece with sweet release
  The load dropped at my feet.

With empty hands on foreign lands—
  You chose that we would meet.

My feet fell stock on a steady rock
  and I began my anxious ascent.

The closer I came, the more I felt shame.
 I feared Your swift relent.

Yet as I reached the top, knees only to drop,
 You bathed me in glorious light.

My senses unhindered, eyes opened, heart tendered,
 Oh, what a beautiful sight!

"Just touch the blues, and smell the greens,
 and warm yourself in yellows."

"Savor the still, or dance if you will,
 in the shade of My umbrella!"

You spoke the Word like I've never heard,
 My spirit expelled such surprise . . .

The You I had known, had been gently shown
 Was transfigured before my eyes!

"Oh, Sweet Lord," my heart adored,
 "I beg You, please let me stay!"

"Come here, My child, let Me hold you a while,
 For you must return today."

He extended His hand so that I'd gently land
 On the path where I belong.

But refusing to retreat, I chose to leap
 Falling from a cliff headlong.

He heard me weep, set me on my feet
 kissing elbows I'd skinned and burned.

Saying, "There's only one reason for your mountaintop
season—
Take to the valley all you have learned."

"But the pain's too intense, the noise immense,
I'd rather be up where You are!"

"That's the captive's cry," my Lord replied,
"the lost are down where you are!"

"Don't despair, I'm as close as a prayer . . .
I'll always descend to you."

"Let's fellowship sweet 'til your mansion's complete,
In the meantime, the mountains are few."

Though that time's come and gone, and life just goes on,
I'll never be quite the same

Since that day I remember, when all else surrendered

. . . and Jehovah called out my name.

The Treasury

We beg and plead and moan and cry
To make sense of this place.
We sweat and strive to fix it all
Then seek You just in case.
When will we learn to listen
To the Master clearly say,
"Seek treasures tucked into My Word . . .
It's there you'll see My face!"

Oh, hasten that sweet moment
When we'll know as we've been known.
Such secrets of Your glory
Cannot be grasped, they're shown!
This fleeting puzzle makes no sense
Except in You alone
And missing pieces swell our faith
And stretch us 'til we've grown.

The Company
I Keep

Let me be known by the company I keep
By the One who determines each day that I greet
From the moment I wake 'til He rocks me to sleep
Let me be known by the company I keep!

Let me be known by the company I keep
When the valleys are low and the mountains are steep
By the One who holds fast when swift waters are deep
Let me be known by the company I keep!

Let me be known by the company I keep
By the One who implores me to sit at His feet
And quickens my soul to discern what is deep
Let me be known by the company I keep!

Let me be known by the company I keep
Eclipsed by Your presence that I may decrease
'Til all You have chosen this traveler to meet
No longer see me but the Company I keep.

A Momentary Rapture

The anguish of my humanness
Extends a ready plea
But for a brief encounter
Would You set my spirit free
To glide where words are meaningless
And earth's mere shadows vain
Where flesh is uninvited
And self is fully drained?

Please rapture me in spirit
Let me ride on wings above
To taste Your glory's sweet unknown
And bask in perfect love . . .
Where praise is not requested
But my soul's compulsory.
Draw me near Your presence,
Lord, disclose Yourself to me.

Let me rise above the stocks and chains
This body guarantees
Shaking loose the cowardice
Of my conformity.

Call me, Father, summon me
To places never meant
To leave a visitor quite the same
When his sweet time is spent.

"Send forth your light and your truth,
let them guide me; let them bring me to your
holy mountain, to the place where you dwell.
Then will I go to the altar of God, to God,
my joy and my delight."
PSALM 43:3–4

The Mountain

I hear You call, "Come meet with Me,
See that which eyes are veiled to see
And ears in vain will please to hear
Except your heart should draw you here!"

"Abandon deeds down at My feet
No crowns as yet, no Judgment Seat
Surrender here all noble plans
Just lift to Me your empty hands."

"I'll fill them with the richest fare
And circumcise your heart to dare
To reach beyond all earthliness
For on this mount you're Mine to bless."

"With gentle hands 'til heart is stretched
I'll have My Word upon it etched
'Til only knees can catch the fall
Of one who cries, 'You are my all!' "

Seasons

Simple things . . .

    Like a brisk catch of breath
        in the first Autumn breeze,
Like pale pink buttercups
      and a Springtime sneeze,
        Like pink children's cheeks
      on the beach at Summer's play,
    Like praying it might snow
        this year on Christmas Day. . . .

Remind me that some things never change

    Like Your unfailing love.

"As long as the earth endures,
    seedtime and harvest,
      cold and heat,
    summer and winter,
      day and night
    will never cease. . . .

    [For] Jesus Christ is the same yesterday and
today and forever."
GENESIS 8:22; HEBREWS 13:8

Heaven

How infinite Your grace for us
As You've prepared a place
Beyond our mere imaginings
With no familiar trace.
We'll find our custom-built abode
Beyond the pearly gate,
Step-by-step down streets of gold
Oh, I can hardly wait!

Yet, Lord, I feel I must express
What Heaven means to me.
It's not the mansion of my dreams
That I receive for free.
You see, Lord, if You'd given me
A hut beneath a tree
Or led me to a country shack
Where I'd forever be

I'd make my home there happily
For always and a day
If You'd make just one promise
You, too, would come and stay.
Because, My Lord, one thing is sure
So search my heart and see
It's not reward my heart leaps for—
You are Heaven to me.

Excellence

I am not about to suggest that either of the next two poems was inspired by tender moments with God. I would like to suggest, however, that sometimes life would be a little easier to take if we learned to steal a few moments out of every day simply to lighten up.

For today's frustrated woman, trying to do all she can do (rather than be all she can be), here's Biblical permission to resign a few activities—

> "Whatever you do, do it all for the glory of God."
>
> 1 Corinthians 10:31

Let's face it. None of us can do a thousand things to the glory of God. And in our vain attempt we stand the risk of forfeiting a precious thing—EXCELLENCE. Oh, that we might discern the will of God, surrender to His calling, resign the masses of activities, and sell out to do a few things well. What a legacy that would be for our children.

> "And this I pray . . . that ye may approve things that are excellent."
>
> Philippians 1:9–10 KJV

Nowhere on earth is a woman's role more distorted than daytime television. This selection has been written in response to her example.

Superwoman's Freedom Plea

Oh, Lord, who said there's just One Life to Live?
I'm sure I'm livin' a thousand!
The few times I do awake to pray
All My Children start arousin'!

Uh-oh! No time for quiet now
Think quick! The day's beginnin'!
I'll try to recall all Oprah's advice. . . .
Then my head starts spinnin'—

Make those younguns religious,
cautious but not suspicious
And watch their self-esteem!
Yet you be professional, look sensational
And keep that house squeaky clean!

And perish the thought you'd forget the needs
Of that marvelous man you married
Why, throw yourself before him
when he raises his eyebrows
And quit thinking, "I'd rather be buried!"

Oops, now I'm late for work, the kids hate their clothes
And the baby's got a cough
As the World Turns so quickly, I'm severely tempted
To take the next jump off.

Surely they're kiddin', Is there anyone left
Who's honestly Young and Restless?
As for me, I feel centuries old, completely worn-out
And cellulite infested!

It's gonna take more than Ryan's Hope
for this woman to survive.
I cannot abide another deep breath of these
Days of our Lives!
Superwoman? She's a curse. To fake her is impossible!
And if I try for one more day, I'll wind up in
General Hospital!

I've gotta be here, I've gotta be there
I frankly cannot face it.
Rescue me from havoc, please, show me what is basic!
Slow me down, Lord, save this life and
keep my eyes on You.
Satan can have this rat-race world—

Thank God, I'm just passin' through.

Servant

Few of us are really confused about the issue of servanthood. Christ made Himself crystal clear when He said:

> "Whoever wants to become great among you must be your servant, and whoever wants to be first must be slave of all. For even the Son of Man did not come to be served, but to serve, and to give his life as a ransom for many."
>
> MARK 10:43-45

For most of us the problem is not serving. The problem is that inherent in the title *Servant* is the inevitable *Servee*. These words are dedicated to every leader who has ever wished that just one time someone would shut up and follow.

The Servant's Plea

Lord, I could be much better
As You would surely see
If I didn't have the folks to serve
That You have given me.
Yes, I could keep my halo straight
Above my little head
If coaching them to action
Wasn't moving tons of lead.
I would be so righteous
I would live my life in prayer
If You would kindly tell this group
To get out of my hair.
I would be so spiritual—
A Christian so much finer
If You would simply give to me
A flock without a whiner.
I wouldn't know how to behave
Without a tear to wipe
Or the chance to hurry to my church
And hear those people gripe.
Who cares on Sunday mornings
If there's donuts for the mob?

Give them something new to think
Or get those folks a job!
Yes, Lord, it's true, I'd love to serve
Beneath that lovely steeple
If You would take a good, stiff broom
And sweep out all those people.
What, Lord, is that You up there?
What's that I heard You say?

I said,
Your righteous robe is in a twist—
I'll untangle if I may.
It's clear you missed the point
While looking pious on your pew.
It's not what you can do for them
But what they'll do for you.
You see, My Child, some years ago
I signed this guarantee
To make you look much less like you
And far more just like me.
Take a closer look at them—
You'll see they are a gift
I'll use to reap your harvest, Child,
As grain from chaff I'll sift
And make you free to love them
As you have been so loved
So, go ahead, serve with a smile—
They're airmail from above!

Vulnerability

Without a doubt, the strongest emotion I have ever experienced has been in response to my children. There is little debate that motherhood takes a woman to the "breadth, length, depth, and height" of the human psyche.

On a summer vacation in 1991, I sat perched on a large rock watching my husband and children prepare to take on the Texas-famous rapids of New Braunfels. Our three children, then eleven, nine, and five, all hooked up in inner tubes like little ducklings behind their father. Their bare feet were secured as tightly around the tummy of the one in front of them as they could wind them.

The "shoot" seemed to jump up and grab them as they began their frantic ride through the rapids. As they dipped, swirled, and spit water from their mouths, their expressions were so priceless that I laughed until my side screamed. Without warning the laughter suddenly transformed into tears. Not just any kind of tears. Overwhelming, uncontrollable tears. The attention-getting kind. As people stared at me, I thought of faking a heart attack, but I was afraid someone would try to give me mouth-to-mouth.

I was utterly humiliated, but I was not confused. I knew exactly what had done it. I was dramatically and painfully confronted by the incarnation of my own vulnerability. There was my Achilles' heel . . . fashioned by the four people

I love more than any others in the whole wide world. My life could be changed in the twinkling of an eye over any one of those creatures. In a second, my overwhelming love for them was transformed into excruciating vulnerability.

I cannot relate to any part of my Savior's perfection or His agony on the cross. The closest I can crawl into the backdrop of that pivotal scene in history is to imagine the emotions of that tender, naive mother. What must her thoughts have been as she stood below her own suspended vulnerability? I doubt this could have been a woman who stood in stoic control of her silent pain. Why would her Son have been so moved in the midst of His own ripping war with death that He cried out to His friend, "John, take care of her! Hold her! Lift her collapsing frame and assure her you'll be there for her! Do something!" Somehow I feel rather certain that this mother was just like the rest of us. She cried out for her own death to escape the intolerable sight of His. Suffering to hide her own eyes yet maternally unable to do so. She couldn't bear to stay, yet she couldn't bear to leave. How different these moments must have been to "ponder" than those surrounding His birth. How she must have tried to shake the pictures out of her head for years to come. How she must have awakened and felt for just a second that it was all a bad dream. And she'd find Him back in that manger. Safe and sound. But it was no dream. It was a nightmare.

Since having my own babies, I have often wondered how different things might have been if God, in His perfect sovereignty, had allowed Mary to know the fate of her first-born Son. The stirring thoughts that run through my mind are simply summed—thank God it wasn't me.

A Mother's Thought at Christmas

Had Mary known, just she alone, when in her arms a baby lay
The pain and sorrow of His tomorrow, sin in its ultimate
    display,
Would she have hidden Him and safely bidden Him
    and quickly run for His life?
Or could she have faced with no attempt to replace
    His inevitable appointment with strife?

What if she had known, through a vision been shown,
    the fate of His downy soft head
Which her cheek brushed gently as He cooed so contently,
    absent all feeling of dread
Of a day far too soon, the sun peaked at noon
    when men filled with hatred and scorn
Would puncture His skin and abruptly press in
    a crown protruding with thorns?

Had Mary known all along the fate of the palm
    she uncurled carefully with her thumb
The hideous sound that a hammer would pound
    when to a nail His palm would succumb

Would her grasp have grown tight as she clutched with her
    might each tiny, searching finger
That would stretch out in pain, no relief to be gained
    as the minutes 'til death only lingered?

What if Mary had perceived the message received
    in the swaddling clothes wrapped 'round Him
That they only foretold a body grown cold
    and the grave clothes that eventually bound Him?
And the clothes He'd wear from His body they'd tear,
    each garment from the other
As they cast their lots no mercy is sought.
    An eyewitness you'll be, Dear Mother.

As my mind still wanders over that one who pondered
    each moment in that stable
If she had known what Scripture has shown,
    would she have changed it if she were able?
I realize now as my knees drop to bow
    something of the God of Glory.
Had He told her these things, what Christ's future would bring,
    He would have told her the rest of the story—

"Yes, Dear One, who holds my Son,
    lifting Him from a hard, wooden manger,
He'll be a man of sorrows, all grief to borrow,
    from birth He'll be in danger.

"On a tree replete with sin's defeat
    He'll soon die in your very own stead.

No earthly throne, He'll die alone, and thorns will crown His
    head."

"Grieve only a while o'er the loss of My Child,
    God incarnate in this baby boy.
The grave will soon see the captives set free
    and your heartache will turn to joy!"
The angels restate, 'How long will You wait
    to give Him all You've longed for?'
My patient reply, First He must die . . . His grave is the Open
    Door!"

"As life came from the womb, there's life from the tomb.
    My plan is being perfected.
There's a place I prepare after sin I repair,
    for My children, My heart's own Elected . . .
Where all bow at His feet, death in defeat,
    and call Him the Lord of all lords!
Blessed choruses ring, 'He's the King of all kings!
    His Word a double-edged sword!' "

"For now, My child, but for a while, cuddle Him all you can.
Gather hay from the loft, sing a lullaby soft, 'Sleep, Baby,
    Blessed God-man.'
So much work must we do when time becomes due.
    Rest for now, My Darling, don't cry.
Stars, shine bright! Dance on His face tonight!
    Look up, your redemption is nigh!"

He is God's Son, the Only One through Whom men can be
restored.

Dry your tears, incline your ears. Your pain is not
ignored.

Hail His Majesty, the Prince of Peace, the Bright and
Morning Star,

Bow each knee, and tongues proceed, praise Him wherever
you are!

Fresh Prints

We're inundated with the news
That all is at unrest.
We've not a clue
What this world's coming to,
Just thank the Lord we're blessed.

Beloved, this very day
You thought you'd never live to see
Is just the one God preordained
And chose for you and me.

We're not called to shake our heads
And utter "what a pity."
We're called as candles on a hill
And towers in the city.

We can draw far more to Christ than tracts
Or fancy steeples
We are proof in breathing flesh—
God moves among His people!

Please understand, this race you run
Is not just for your prize.
Grab young hands, courageous band,
Run for their very lives!

For us, we must live for today,
For them, live for tomorrow.
Redeem the time for many blind
For there is none to borrow!

The prints of history's heroes
Will soon fade into the dust,
If there will be fresh prints, my friend,
It is up to us.

Footprints that walk the talk that says,
"I'll go where You will lead!"
Kneeprints that bridge the gap
And make the hedge to intercede.

God, kick us off our cushioned seats
Don't let us turn our heads!
Let's cease to hide behind the cross
And carry it instead!

You beckon us, "My warriors,
The time has come, ARISE!
Draw your swords, fight the fight,
Sound the battle cry."

"Where are My few who dare to say,
'Come follow Him with me?'
Would you lay down your own dear life
So that My Son they'll see?"

"Consider, Child, carefully—
Am I quite worth the cost?

To surrender hearts to holiness
And count all gains but loss?"

"I call you from your comfort zone,
Dare you be one of few?
If you'll not leave fresh prints, My child,
Then I must ask you, who?"

If you'll not lead the way, My child,
Then look around you,

Who?

Steal Away

Steal me away like a child at play
To the fields that flow at Your feet
Spread the grass like a blanket cast
A picnic where two shall eat.

For just a time let joy sublime
Serve our savory course
No fare of woes, no need for no's
Full without remorse.

Animal clouds to guess out loud
Laughing at Your sky
You and me and I shall be
The apple of Your eye.

Time comes back and I must sack
What's left of fish and loaves
Some will taste with hungry haste
Steal away, steal away those!

Things That Remain

Faith—Knowing He can whether or not He does.

Hope—Knowing He will whether or not He has.

Love—Knowing He died whether or not we live.

Delighting thyself in the Lord is the sudden realization that He has become the desire of your heart.

Enemy

You have no power over me

> but what I give you

no authority

> but what I bid you

no voice

> but what I hear

no leading

> but what I follow

no savvy

> but what I ascribe

no hold

> but what I cling to

no skill

    but what I admire

no room

    but what I make you

no gain

    but what I hand you.

You have

    no right.

My Every One

Lead me, Gentle Shepherd
Save me, Lamb of God
Feed me, Bread of Heaven
Alone on paths You trod.

Hear me, Intercessor
Answer, Living Word
Rescue, O Deliverer
Still the waters stirred.

Plead for me, my Advocate
Set me free, O Truth
Soothe me, Tender Comforter
Shake hell's kindred loose.

Doctor, Great Physician
Seek me, Blindless Sight
Grant me, Freely Giver
Usher forth, Dear Light.

Empower me, O Mighty One
Quiet me, my Peace
Keep me, Blest Assurance
From graven hand's release.

Chasten me, my Father
Gracefully restore
Build my house right next to You
Escort me home, O Door.

Echoes From the Pit

Come quickly, O my Caravan
The pit is dark and deep
My head is dizzy from the fall
My way is much too steep.

Clouds obscure the fainting sun
Let night not catch me here
The well echoes my pounding heart
O, Caravan, draw near!

I stretch to listen . . . is that You
The rumbling earth restates?
Hasten please, O Caravan
Dark angels will not wait.

I see the shadow of Your face
I feel Your knotted rope
Blessed sight, O Caravan!
Lifting me with Hope.

Despised

My plane had arrived in Oklahoma City early that afternoon, and I was not speaking until later that evening. I decided to take advantage of the quiet hours and study for my next Sunday School lesson. The text was Isaiah 53. I had been acquainted with the chapter since childhood and had committed it to memory as a young adult, yet in the moments that followed, God allowed me to approach these words with a total freshness of Spirit. I meditated over the familiar passages that so perfectly described the suffering of my Savior. He was . . .

> "despised and rejected of men." (v. 3)
> "afflicted." (v. 7)
> "brought as a lamb to the slaughter." (v. 7)
> "cut off out of the land of the living." (v. 8)

The descriptions stung my eyes as if I had seen them for the first time. A picture so different from the one I usually imagine—The Great High Priest sitting truimphantly at the right hand of the Father interrupting my feeble petitions with power. It seemed so unfitting. The shame He had endured was nearly overwhelming to me in that moment. The prophet Isaiah was inspired by God to pen the definitive chapter on the doctrine of Salvation. One which answers so vividly the question of any God-fearing believer—"Is God satisfied with me?"

> "When thou shalt make his soul an offering
> for sin. . . . He shall see the travail of his soul
> and shall be satisfied." (vv. 10–11 KJV)

The issue is not God's satisfaction with man. God is satisfied with Jesus. One by one the descriptions moved me, but a single verse continued to steal me away with peculiarity. Maybe it was the woman in me. An inborn weakness for romance. An ongoing battle with vanity. I'll never know for sure; nevertheless, my eyes returned to those words over and over which described the Incarnation . . . the very "fullness of the Godhead bodily" (Col. 2:9)—

> "He hath no form nor comeliness; and
> when we shall see him, there is no beauty
> that we should desire him." (v. 2)

No beauty? How can that be? My mind recaptured the familiar pictures on the walls of my fourth-grade Sunday School class. His face looked so kind. His eyes so tender. His mouth turned up at the edges. His hair looked so soft. The pictures themselves had drawn me to Him. And to have seen Him in Person? How could He not have been beautiful? Yet that's what Scripture said—to the human eye there was nothing beautiful to behold.

I sat back in my chair and stared out the window, imagining what it might have been like to have lived when He lived. And to have looked for myself. What would it have been like to have known Jesus personally? To have been the woman at the well. Or Mary, the sister of Lazarus, leaving Martha with the dishes. Can you imagine being the one about whom He said, "He that is without sin among you cast the first stone at her" (John 8:7)? Or the one who took the precious feet that would soon be pierced, anointed them and wiped them with her hair? Just imagine what it might have been like to have lived in a city called Jerusalem and to have known a man they called Jesus the Nazarene.

Beautiful

You know, I've never been a person who saw things like anyone else. I often saw cruelty in the "harmless" games of children and poverty in the paths of the rich. I saw a joyous parade in the Kidron Brook as the water rushed the pebbles clean. So perhaps I saw a different man than most. For You were beautiful to me.

Maybe it was Your eyes. Those eyes that showered fullest attention upon whomever You encountered. Eyes that fastened with such focus, I would have run . . . if I could've moved. Eyes that I realized knew everything there was to know about me. But eyes that reassured, "It is I, be not afraid." There was definitely something about those eyes. While the rest of us bowed our heads to pray, You lifted Yours straight to the Heavens, as if You could penetrate the endless blue and gaze upon the very Throne of God! But, then, You could, couldn't You? And when mine locked with Yours, I glanced into the very soul of God! Those eyes. They were so beautiful . . . so perfect . . . to me.

And those hands—how in the world could a carpenter's callused hands be so soft? Hands which could carve the most exquisite craftsmanship, yet hands which could swing an endless line of giggling children running into Your arms after school. Those precious hands which broke the bread also calmed the raging sea. And to be touched by those hands? Oh, to be touched was to be healed! How I wish I could erase the memory of those hands covered with blood. How I had hoped those scars would be gone when I saw You again. Some scars are too deep to fade, I suppose.

143

And Your heart. Oh, I saw that heart in everything You did! The expressions on Your face. The laughter from Your lips. A heart from which You shouted in the midst of Your own agony, "John, take care of my Mama for me!" A heart which grieved the separation of loved ones. A heart which both hated and cherished the day when the grave would be overcome. Yours was a heart that pounded a passion for hurting people. A heart which ceased that the wages of sin might also cease. Your tender heart. It was so beautiful to me.

And who could forget that voice? The one which spoke the very worlds into being . . . and commanded the light to shine in the darkness . . . then became that light. The voice that thundered, "Lazarus, come forth!" The same voice that spoke the very imaginations of my mind and hidden sins of my heart . . . loudly enough for me to confront my own poverty . . . yet softly enough to spare my dignity before my peers. Simply, "Go and sin no more." The voice that tenderly called out my name. How I long to hear it again . . . That perfect voice which will someday repeat the invitation of a lifetime, "Come and follow me."

Yes, someday . . . maybe someday soon . . . the skies will roll back like a scroll, and the angels will announce, "Hail, His Majesty, the King of kings!" And there You will be! In all Your glory! With unspeakable splendor! And all will adore You! And with certainty, the heart of every man, of every woman, of every child will fellowship in a compulsory refrain, "ISN'T HE BEAUTIFUL!" And, I, with no words to express, will joyfully . . . endlessly agree.

Perhaps love is blind. On the other hand, perhaps only love can truly see. Because, my Christ, my Hope, I want you to know . . . You were always beautiful to me.

The Journal

I came across an old journal just the other day
Written twenty years ago—words that I had prayed
I sat cross-legged on the floor and read its every thought
It chronicled my dreams and dares and battles I had fought.

Ink filled the pages with beginnings' dangled ends
College courses, roommate choices, countless boyfriends
I couldn't help but laugh at my astute ability
To transform such a simple life into complexity.

You'd have thought life's most important theme was know-
    ing if or not
I'd get to know that brand-new guy or keep the one I've got
Little seemed consistent in the life of one who wrote
Except the repetition of, "Oh, God, I love you so."

How on Heaven's Earth could I have known what true love
    was?
Surely youth obscures the wondrous works the Most High
    does
I held the journal to me and I thought of where I'd been
I sighed and cried to wonder,

    "And I thought I loved you then."

As if by a greater force I walked into the other room
Pulled out another journal—prayers weaved from a later
    loom
Ink filled the pages with beginnings' dangled ends
Cutting teeth, sleepless nights, and sticks from diaper pins.

"I think she said 'Mama' today—I know that's what I heard!"
The moments seemed to slip away as quick as they'd occurred
One day I was the Queen of Hearts, the next I was in crisis
Monday my man was worth pure gold, Tuesday I'd quote his
    vices.
Little seemed consistent in the life of one who wrote
Except the repetition of, "Oh, God, I love you so."

How on Heaven's Earth could I have known what true love
    was?
Surely weary moms miss wondrous works the Most High
    does!
I held the journal to me and I thought of where I'd been
I sighed and cried to wonder,

    "And I thought I loved you then."

My hand reached for another in a more familiar place
Written by a woman with a far less youthful face
Ink filled the pages with beginnings' dangled ends
My oldest nearing college and my youngest turning ten.

This one struck my heart, its pages twisting as they turned
With lessons in the making and a few so harshly learned
My prayers were often moans and my questions painful cries
So many sick and hurting—time injures as it flies.

Relief would show up just in time through children's victories
Or sudden understanding of the ancient mysteries
Still little seemed consistent in the life of one who wrote
Except the repetition of, "Oh, God, I love you so."

No longer could I overlook the sole familiar thread
Quilting twenty years of ones—a single sentence said
How could I use those same old words to pen my fickle heart
Through ceaseless ups and downs and a thousand second
     starts?

How on Heaven's Earth could I have known what true love
     was?
I've only just begun to see the wondrous works He does!
His certain words fell on my ear as gently as a song:
"Don't you see, My child? It was true love all along."

"Things were not so simple when you were just a youth
Your words told me the easy things, your heart told me the
     truth
And what more needful time for prayer than when a little
     one
Learns to mock your every move and bedtime never comes?"

"You needed me as badly then as you could need me now
The greater and the simpler both to train your knees to bow
Your words may seem the very same as those right from the
     start
But they fall afresh upon me when I see a different heart!"

"How on Heaven's Earth could you have known what true
     love was?

I met you at each step—for that's the work the Most High
    does
The times I patch and stretch your heart until it's nearly sore
Deepen you to say those words and mean them all the more."

"As surely as the ones which passed—we've miles left to go
Despite your inconsistencies, Oh, child, I'll love you so
You'll one day hold this journal, too, and think of where
    you've been
Ink-filled pages tying those beginnings to their ends
You'll find those same old words—they'll puzzle you again
Lined face will sigh and wonder,

    'And I thought I loved you then.' "

After things pondered . . .
the dreams of a child,
the realities of an adult,
one thing remains . . .

Hope

I've grown old enough to know
That fairies don't have tails
That good men often suffer
While evil men prevail.
I've tried to find that white frame house
With matching picket fences
But found instead black picket signs
And hatred's thorny fences.
I've lived enough of life to see
The innocent maligned
And I've concluded fairness is
A rarity to find.
I've seen the noble dreams of man
Be in an instant shattered
I sigh to see another woman
Used and bruised and battered.
I've seen shots of tiny orphans
As rulers rise and fall
I've stood by stricken parents
And caskets way too small.
I've abandoned childish notions
That life is like pretend
I've tossed paper to the ground and sobbed,
"When will this madness end?"

But I've never grown up quite enough
To leave my hope behind
I'll think I've turned my back on hope
Then bump into the kind
Of Gentle Traveler sent to bind
My wounded faith with love
Who sets my feet upon a Rock
And mind on things above.
Then I find myself still hoping
Old folks won't be left alone
And can't seem to quit believing
Daddies still might move back home.
And that an orphan might just find
A reason to survive
And parents of the missing
Might just find their son alive.
No, I've never grown up quite enough
To scorn sweet signs of Spring
Nor can I help but think a tree
Is happy with a swing.
And you must pardon if I hope
The Pearl of Heaven's Gate
Is the treasure I've adored
And longed to celebrate.
I hope to hug the ones I've loved
And jump on cotton clouds
Where angels sing His holiness
And saints can laugh out loud.
Some bedtime tales are worth the tell—
May one be quickly due

Let Gabriel groom that great white horse
And board Faithful and True.
So let this world's prince mock and scorn
My hope is not ashamed
For in the King of kingdom's grand
My Hope has found a Name.